Aldo Maceri

Fluid Dynamics

Independently Published

Prof. Ing. Aldo Maceri
Past Full Professor of Scienza delle Costruzioni
University of Roma "Roma Tre"
Department of Engineering
Italy

Aldo Maceri - *Fluid Dynamics*

PREFACE

The problem we are examining here is that of building a mathematical model that simulates the behavior of a *fluid*, which can be a gas or a liquid. In this analysis, *Thermodynamics* joins *Mechanics* with a greater role than that usually performed in the case of *solids*. We immediately point out that, even in the presence of thermal effects, the analysis of the main problems of engineering interest can be conducted by simulating the body with a continuous medium. However, we specify that in some engineering applications of *Fluid Dynamics* we encounter fluids for which the continuum model is not admissible (as in the case of *rarefied gases*).

In fluid dynamics, as a rule, non-negligible *mechanical energy dissipation* phenomena occur, with consequent production of *entropy*. If entropy is produced during a transformation, this cannot be canceled and it follows that the transformation considered is *irreversible*. It is therefore obligatory, in order to construct an adequate model of the problem, to resort to the *Thermodynamics of irreversible processes*. However, let us premise some references to *classical Thermodynamics*, based on a theory formulated with postulates.

This book is an introduction to *Fluid Dynamics*. It deduces its fundamental results favoring (when possible) considerations of a *physical nature*.

Roma, 06.02.2024   Aldo Maceri

CONTENTS

Chapter 1

Mechanics of the continuum

1.1 Introduction

The problem we will now consider is that of building a mathematical model that simulates the behavior of a fluid, which can be a gas or a liquid. In this analysis, *Thermodynamics* joins *Mechanics* with a greater role than that usually performed in the case of solids. We immediately point out that, even in the presence of thermal effects, the analysis of the main problems of engineering interest can be conducted by simulating the body with a continuous medium. Strictly speaking however, as is well known in the *Structure of matter*, the volume occupied by the atoms present in a body is only a very small part of the volume of the body. For example, in a solid-state piece of steel, the ratio is on the order of 10^{-14}.

It may happen that non-negligible mechanical energy dissipation phenomena occur in the problem under analysis. In this case, in order to construct an adequate mathematical model of the problem, it is necessary to resort to the *Thermodynamics of irreversible processes*.

1.2 Review of classical Thermodynamics

We call *system* a portion of space occupied by matter, *environment* the rest of the universe. We denote with $V(t)$ the portion of the three-dimensional space occupied by the system at the instant $t \in \,]t_i, t_f[$ and with $\Sigma(t)$ its *boundary* (*i.e.* the *surface* of the system). The symbol t_i [resp. t_f] denotes the *initial instant* [resp. *final instant*] of the time interval in which we analyze the phenomenon.

We formulate a theory of *classical Thermodynamics* via postulates. We assume as *first postulate*

[1.1] *A system can exchange mass and energy with its environment.* ■

In [1.1] *mass* must be understood in its various forms (*i.e.* in its various states of aggregation) and *energy* in its various forms (potential, kinetic, electromagnetic, etc.). The postulate [1.1] states that everything that the system exchanges with the environment is associated with an exchange of mass and/or an exchange of energy. For example, the exchange of *mechanical momentum* is associated with an exchange of mass; the exchange of *electromagnetic momentum* is associated with an exchange of energy.

Clearly this first postulate of *Thermodynamics* is valid only outside the relativistic field. In fact, in the relativistic field it is not possible to exchange mass and energy independently.

Of course, admitting this first postulate of *Thermodynamics* implies that the system can exchange with the environment either energy but not mass or mass and energy. In fact, the exchanged mass brings with itself (even in a non-relativistic field) its energy.

A quantity is said to be *extensive* [resp. *intensive*] when its numerical value depends on [resp. does not depend on] the extension of the system.

Evidently if G is an *extensive quantity*, called g the *specific quantity* (also called *density* of G), it results

$$G = \int_V g \; dV \, .$$

We call *punctual flow* (or *local flow*) at the instant $t \in \,]t_i, t_f[$ of an extensive quantity G through a surface $\Lambda(t)$ at a point $P(t)$ (of $\Lambda(t)$), and denote by the symbol φ_G, the quantity of G which crosses in the unit of time a unit of area of $\Lambda(t)$ to which $P(t)$ belongs. The surface $\Lambda(t)$ is supposed to be *regular*, so that each of its points admits a tangent plane (and a normal $\boldsymbol{n}$). Evidently

$$\varphi_G = g \; \boldsymbol{v} \times \boldsymbol{n} \, .$$

We call *flow* (or *global flow*) at the instant $t \in \,]t_i, t_f[$ of an *extensive quantity* G through a surface $\Lambda(t)$, and denote by the symbol $\Phi_G(t)$, the real number

$$\Phi_G(t) = \int_{\Lambda(t)} \varphi_G \, d\sigma = \int_{\Lambda(t)} g \, \boldsymbol{v} \times \boldsymbol{n} \, d\sigma \,.$$

We call *punctual production* (or *local production*) at the instant $t \in \left]t_i, t_f\right[$ of an *extensive quantity* G in a volume $V(t)$ at a point $P(t)$ (of $V(t)$), and we denote by symbol δ_G , the quantity of G which is created (or destroyed) in the unit of time in a unit of volume (of $V(t)$) to which $P(t)$ belongs.

We call *production* (or *global production*) at the instant $t \in \left]t_i, t_f\right[$ of an *extensive quantity* G in a volume $V(t)$, and denote by the symbol $\Delta_G(t)$, the real number

$$\Delta_G(t) = \int_{V(t)} \delta_G \, dV \,.$$

A system is said to be in *thermodynamic equilibrium* if, at each of its points (*interior* or *boundary*), all local flows and all local productions are zero.

Let us now formulate the *second postulate* (which is usually called the *first law of Thermodynamics*).

[1.2]   *There exists an extensive quantity U (which we call internal energy) and an extensive quantity S (which we call entropy) each of which is a function (in conditions of thermodynamic equilibrium) of a finite number of extensive quantities of the system, which we call state variables.* ∎

The postulate [1.2] allows us to assume S [resp. U] as state variable. Therefore, called $X_1, ..., X_m$ ($m \in N$) the other state variables from which U [resp. S] depends, we have

$$U = U(S, X_1, ..., X_m)$$
$$[\text{resp. } S = S(U, X_1, ..., X_m)] \, .$$

We also say that the system has the $m + 1$ degrees of freedom $S, X_1, ..., X_m$ [resp. $U, X_1, ..., X_m$].

We postulate now

[1.3]   *The functions U and S are continuous, homogeneous of the first degree and it results*

$$\frac{\partial U}{\partial S} > 0 \ , \quad \frac{\partial S}{\partial U} > 0 \ .$$

Furthermore, in conditions of thermodynamic equilibrium, called T [1.1] *the absolute temperature of the system, it results*

(1.1) $$\frac{\partial U}{\partial S} = T \ , \quad \frac{\partial S}{\partial U} = \frac{1}{T} \ . \ \blacksquare$$

We explicitly note that, for the postulate [1.3], we have, $\forall \lambda \in \Re$

$$U(\lambda S, \lambda X_1, \dots, \lambda X_m) = \lambda U(S, X_1, \dots, X_m)$$
$$S(\lambda U, \lambda X_1, \dots, \lambda X_m) = \lambda S(U, X_1, \dots, X_m) \ .$$

[1.1] Also called *thermodynamic*. The *absolute temperature scale* was established by *Lord Kelvin*, who assigned $T=273,16\ °K$ to water and ice in conditions of thermodynamic equilibrium (at a pressure of 1 atm).

Let us now formulate the *fourth postulate* (which is usually called the *second law of Thermodynamics*)

[1.4] *Internal energy U can be produced or destroyed. Entropy S can be produced but not destroyed* [1.2]. ∎

We now postulate (*third law of Thermodynamics or Nernst's principle*):

[1.5] *When T tends to zero S tends to zero.* ∎

We now denote with *E* the *total energy* of the system (in all its forms) and we postulate that (*principle of conservation of energy*)

[1.6] *If a system is in thermodynamic equilibrium it results* $\Delta_E = 0$. ∎

[1.2] The second law of Thermodynamics also admits other equivalent formulations. Note that it is not possible to attach physical meaning to entropy. However, its importance is such that it is necessary to get used to considering it as an attribute of the system, just like mass or volume.

We call *transformation* of the system any evolution of the system that takes it from a state of thermodynamic equilibrium to another state of thermodynamic equilibrium. If the final state coincides with the initial one, the transformation is called *closed* or rather *cycle*. If the evolution of the system does not lead to a state of thermodynamic equilibrium or does not start from a state of thermodynamic equilibrium, it is preferable to speak of a *process* rather than a *transformation*.

A transformation is said to be *reversible* if it is possible to bring the system and the environment back to their initial conditions.

Generally, a cycle brings the system back to the initial conditions but not the environment. Therefore, it is reversible if it consists of two reversible transformations.

If entropy is produced during a transformation, this cannot be canceled and it follows that the transformation considered is *irreversible*.

A transformation (or process) is said to be *adiabatic* if (called Q the flow of energy in the form of heat) is $Q = 0$.

A transformation (or process) is said to be *isentropic* if

$$\frac{dS}{dt} = 0 .$$

We consider a system in thermodynamic equilibrium. We know that it is completely characterized by $m \in N$ state variables (in the sense that if the value of these m parameters is known, it is possible to determine the value of any thermodynamic quantity of the system). Therefore, the thermodynamic behavior of the system is fully characterized if m independent equations are known in the state parameters: the fundamental relation and $m - 1$ independent equations of state.

For a gas, consisting of a single chemical phase and a single physical phase, in conditions of thermodynamic equilibrium, the state parameters are the mass M, the volume V, the entropy S (so that $m = 3$). We know that the internal energy state function U exists, so that

$$U = U(S, V, M)$$

so

$$dU = \frac{\partial U}{\partial S}\, dS + \frac{\partial U}{\partial V}\, dV + \frac{\partial U}{\partial M}\, dM \ .$$

We postulated that

$$\frac{\partial U}{\partial S} = T \ ;$$

moreover, since $\frac{\partial U}{\partial V}\, dV$ is an energy,

$$\frac{\partial U}{\partial V}$$

must be a *pressure* and we denote it with the symbol p.

Similarly, $\frac{\partial U}{\partial M}$ is energy per unit mass and we call it the *electrochemical potential* μ .

Therefore, for the system under consideration the fundamental relation is (*Euler*)

(1.2) $\qquad dU = T\,dS + p\,dV + \mu\,dM$

(note that, if the system does not exchange mass, the third addendum to the second side is zero).

If we add to the fundamental relation two equations of state

$$p = p(S, V, M)$$
$$\mu = \mu(S, V, M)$$

we fully identify the thermodynamic behavior of the system.

Normally the fundamental relation is obtained by

hypothesizing a model (*i.e.* conjecturing an expression of it) and then going to verify what its field of validity is.

If a gas has r components (with r positive integer), in perfect analogy we obtain the fundamental equation (*Euler*'s)

$$dS = \frac{1}{T}\, dU + \frac{p}{T}\, dV - \sum_{i=1}^{r} \frac{\mu_i}{T}\, dM_i$$

to which $r + 1$ equations of state must be added (since $m = r + 2$).

The equations of state express the extensive parameters as a function of the intensive parameters. An equation of state of a system composed of m grams of a gas of molecular weight M can be approximated by the law [1.3], setting $N = m/M$

(1.3) $\qquad\qquad pV = N R_0 T$.

A substance which exactly obeys (1.3) is called *perfect gas* or *ideal gas*. All real gases have a behavior which is simulated by (1.3) only approximately.

The second equation of state of an ideal gas is

(1.4) $\qquad\qquad U = c_V T$

where the constant c_V (which is called *specific heat at constant volume*) is equal to $\frac{3}{2} R$ for a monatomic gas,

---

[1.3] In formula (1.3) R_0 denotes a *universal constant* for all gases, equal to $R_0 = 8316,6$ $N \cdot m/°K = 1,986\ Cal/°K$. It is also used to place $R = R_0 /M$ in (1.2.3). Equation (1.3) summarizes *Boyle*'s ($pV = cost$), *Gay–Lussac*'s

$$(V = V_0\left(1+\frac{T}{273,14}\right),\ p = p_0\left(1+\frac{T}{273,14}\right))$$

and *Avogadro*'s (equal volumes of any gas, under equal conditions of temperature and pressure, contain the same number of molecules) laws.

to $\frac{5}{2} R$ for a diatomic gas [1.4].

Moreover, for an ideal gas it is easily obtained that, called c_p the *specific heat at constant pressure*, it results (*Mayer*)

$$c_p - c_V = R .$$

REMARK 1.1 We note that an *adiabatic transformation* of a perfect gas has the equation $p V^{\gamma-1} = \text{const}$ (being $\gamma = \frac{c_p}{c_V}$). In fact, we will see that an *adiabatic* is *isentropic*. Therefore from (1.2), (1.4) follows

$$c_V \, dT = p \, dV$$

and from here and from (1.3) we derive

$$\frac{dT}{T} + \frac{c_p - c_V}{c_V} \frac{dV}{V} = 0$$

from which

[1.4] This result is obtained in a simple way using the *kinetic theory*.

$$\frac{dT}{T} + (\gamma - 1)\frac{dV}{V} = 0 \, ,$$

hence the thesis. ∎

REMARK 1.2 The quotient

$$c = \frac{q}{T_2 - T_1} \, ,$$

where q is the amount of heat required to bring the unit mass of the substance from T_1 to T_2, is called the *specific heat c of a substance between the two (absolute) temperatures T_1 and T_2* .

Thus, for a body weighing P Kg, the quantity of heat necessary to heat it from T_1 to T_2 is

$$Q = cP(T_2 - T_1)$$

Note that the exact definition of c is

$$c = \lim_{T_2 \to T_1} \frac{q}{T_2 - T_1} = \frac{dq}{dT} \quad . \blacksquare$$

In the case of gas, the *specific heat at constant pressure* c_p and the *specific heat at constant volume* c_V are important.

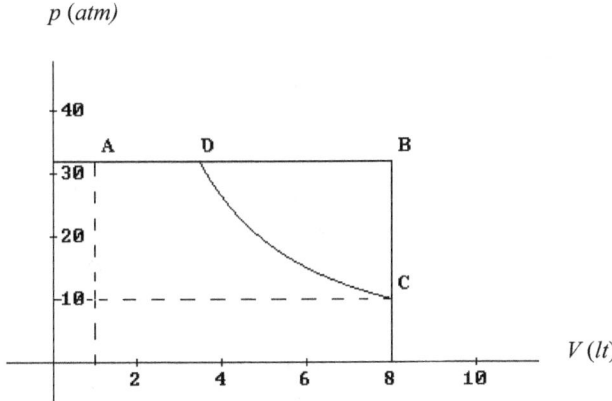

fig. 1.1

PROBLEM 1.1   *A fluid (having $\gamma = 2,4$) evolves according to the transformation $A \to B \to C \to D$ of fig. 1.1, where the transformation $C \to D$ is adiabatic. Determine the work done and the heat exchanged.*

Solution. The transformation $C \rightarrow D$ has the equation $pV^K = const$ (with $K = \gamma - 1 = 1{,}4$).

As for the work, we have (since in $B \rightarrow C$ the volume is constant)

$$L = L_{AB} + L_{BC} + L_{CD}$$

$$= \int_A^B p\, dV + \int_B^C p\, dV + \int_C^D p\, dV$$

$$= p \int_A^B dV + 0 + \int_C^D p_D V_D^K V^{-K}\, dV$$

$$= p(V_B - V_A) + p_D V_D^K \int_C^D V^{-K}\, dV$$

$$= 32 \cdot 7\ atm \cdot lt + p_D V_D^K \left[\frac{V^{-K+1}}{-K+1}\right]_C^D$$

$$= 224\ atm \cdot lt + \frac{p_D V_D^K}{1-K}\left(V_D^{1-K} - V_C^{1-K}\right)$$

$$= 224 + \frac{p_D V_D^K V_D^{1-K} - p_C V_C^K V_C^{1-K}}{1-K}$$

$$= 224 + \frac{p_D V_D - p_C V_C}{1 - K} \ atm \cdot lt.$$

From here (observing that from the $p_C V_C^{1,4} = p_D V_D^{1,4}$ follows $V_D = 2,7 \ lt$) we get

$$L = \frac{32 \cdot 2,7^{1,4} - 10 \cdot 8^{1,4}}{-0,4} = 224 - 176$$

$$= 48 \ atm \cdot lt.$$

As for the heat flux, let us first observe that for a real gas [1.5]

$$c_p = AR \frac{\gamma}{\gamma - 1} \ , \qquad c_V = AR \frac{1}{\gamma - 1}.$$

In the case of air, resulting

$$A = \frac{1}{427} \frac{cal}{Kgm} \ , \qquad R = 29,27 \frac{Kg \ m}{Kg \ ^{\circ}C} \ ,$$

we have

$$c_p = 0,239 \frac{cal}{Kg \ ^{\circ}C} \ , \qquad c_V = 0,171 \frac{cal}{Kg \ ^{\circ}C}.$$

[1.5] For an ideal gas $A = 1$.

Aldo Maceri

Furthermore, using the equation of state (1.3), we obtain

$$T_A = 11 \ °C \ , \ T_B = 87,5 \ °C \ , \ T_C = 27,5 \ °C \ .$$

Thereby

$$Q = Q_{AB} + Q_{BC} + Q_{CD} = Q_{AB} + Q_{BC} + 0$$
$$= c_p(T_B - T_A) + c_V(T_C - T_B)$$
$$= 18,3 \ \frac{cal}{Kg} - 10,3 \ \frac{cal}{Kg} = 8 \ \frac{cal}{Kg} \ . \ \blacksquare$$

PROBLEM 1.2   *Two systems are isolated from the external environment (by impermeable, anergotic and adiabatic walls) and are separated between them by an impermeable, anergotic, diabatic wall. System 1 consists of two moles of He (which is monatomic), system 2 of three moles of O_2. The total energy of the system is 6000 Kcal. Calculate U_1 and U_2 at equilibrium.*

Solution.   At equilibrium $T_1 = T_2$. Since

$$U_1 = \frac{3}{2}\, N_1\, R_0 T_1 = \frac{3}{2} \cdot 2 \cdot 1{,}98 \cdot T_1 \ \ Kcal$$

$$U_2 = \frac{3}{2}\, N_2\, R_0 T_2 = \frac{5}{2} \cdot 2 \cdot 1{,}98 \cdot T_2 \ \ Kcal$$

it must be at equilibrium

$$U_1 + U_2 = 6000 \ \ Kcal$$

$$\frac{U_1}{3 \cdot 1{,}98} = \frac{U_2}{7{,}5 \cdot 1{,}98}$$

and then

$$U_1 = 1720 \ \ Kcal \ , \quad U_2 = 4280 \ \ Kcal \ . \ \blacksquare$$

1.3   The balance equations

Let us consider a quantity G of the system (for example mass or energy), function of the time variable

Aldo Maceri

$t \in \,]t_i, t_f[$ [1.6] . We denote by g the quantity of G contained in the unit volume of the system. Therefore g is a function of the spatial variables x, y, z and of the time variable t and results

$$\forall t \in \,]t_i, t_f[\qquad G(t) = \int_{V(t)} g(x, y, z, t) \; dx \, dy \, dz \; .$$

Evidently G can vary in the time interval $]t, t + dt[$ only if we introduce another quantity of G from the outside or if it is created or destroyed inside the system itself. So, the change of G in the time interval $]t, t + dt[$ is equal to the flow of G (through $\Sigma(t)$) in the time interval $]t, t + dt[$ plus the production (which if negative is a destruction) of G in $V(t)$ in the time interval $]t, t + dt[$. Therefore in $]t, t + dt[$

$$(1.5) \qquad \qquad \frac{dG}{dt} = \Phi_G + \Delta_G \; .$$

[1.6] $t_i \in [0, +\infty[$ is the initial instant; t_f is the final instant (*i.e.* a real number greater than t_i or the symbol $+\infty$).

The (1.5) is called the *balance equation* (*of G*).

The balance equation is also valid for thermodynamic quantities. However, when it is applied to thermodynamic quantities it is necessary to assume that the system is in thermodynamic equilibrium.

A quantity G such that $\Delta_G = 0$ is said to be *conservative*. If in (1.5) G is a conservative quantity, (1.5) is also called the *conservation equation* or *conservation principle*. Evidently in an *isolated system* [1.7] every conservative quantity is constant.

We apply the balance equation to the total energy E of the system, which by postulate [1.6] is a conservative quantity. We have

(1.6) $$\frac{dE}{dt} = \Phi_E \ .$$

[1.7] A system is said to be isolated if $\forall t \in \ ]t_i, t_f[$ the flow across the boundary $\Sigma(t)$ (of the volume $V(t)$ occupied by the system at time t) of any quantity is zero.

We postulated that the system can exchange with the environment either *energy but not mass* or *energy and mass*. We then split Φ_E into the sum of Φ_{EE} (energy flux that is not associated with mass exchange) and Φ_{EM} (energy flux associated with mass exchange).

It is convenient (and it will be clear later why) to consider the energy flow Φ_{EE} sum of an aliquot Q (which we call *heat flow* per unit of time or *thermal power* or simply *heat*) and of an aliquot L (which we call *work flow* per unit of time or *mechanical power* or simply *work* (per unit of time)).

Therefore, if Σ does not allow mass exchange between system and environment, it results

(1.7) $$\frac{dE}{dt} = Q + L \ .$$

In (1.7) (which is one of the forms of the *principle of conservation of energy*) it should be highlighted that Q and L are two forms of *energy flows*, *i.e.* they are both *energies in transit.*

The variation of E in a time interval $[t_1, t_2]$ is obtained by integrating the two sides of (1.7) with respect to time (from t_1 to t_2).

We now carry out the balance of a quantity that is not conserved. A typical quantity of thermodynamics that is not conserved is *entropy*. The internal energy U is also not conserved. But we know something more about entropy: not only is it not conserved but it can't even be destroyed.

We immediately observe that since the quantity we want to balance is a thermodynamic quantity, in order to be able to define it, the system must be in thermodynamic equilibrium. So, from (1.5) we get [1.8] that in $]t_i, t_f[$

[1.8]   In this simplified formulation there is an inconsistency, because in conditions of thermodynamic equilibrium the flows and the productions are zero. However, in the exact setting, which preliminarily gives all the concepts of *equilibrium Thermodynamics*, this inconsistency does not exist.

Aldo Maceri

$$(1.8) \qquad \frac{dS}{dt} - \varPhi_S = \varDelta_S \ .$$

By the postulate [1.4] $\varDelta_S \geq 0$. So, the (1.8) gives

$$(1.9) \qquad \frac{dS}{dt} - \varPhi_S \geq 0 \ .$$

We postulated that the system can exchange with the environment either *energy but not mass* or *mass and energy*. Therefore, as for the total energy, we split $\varPhi_S$ into the sum of an entropy flow $\varPhi_{SM}$ associated with mass exchange and an entropy flow $\varPhi_{SQ}$ associated with energy exchange but not mass exchange. Of course, if Σ does not allow mass exchange then $\varPhi_{SM} = 0$.

We now give a fundamental definition of *Thermodynamics* which defines *heat* and *work a posteriori*. $\varPhi_{SQ}$ is associated with the flow of energy in the form of heat. *Thus, that particular form of energy flow which we have called work does not bring, associated with itself, flow of entropy.* In conclusion, entropy can flow (through Σ) together with mass or together with energy or together with both, but it cannot flow alone (through Σ). So, if the system is isolated it results $\Phi_S = 0$.

We denote by φ_{SQ} the *local flow of entropy* (through Σ) *not associated with mass flow*, so that

$$\Phi_{SQ} = \int_\Sigma \varphi_{SQ} \, d\sigma \ .$$

As we can rigorously demonstrate, if T is the *absolute* (or *thermodynamic*) *temperature* of the system, it results

$$T \, \varphi_{SQ} = q \ .$$

In particular, if T is constant on Σ , we have

$$\Phi_{SQ} = \frac{Q}{T} \ .$$

1.4 *Thermodynamics* of irreversible processes

The *Thermodynamics of reversible processes* (which is also called *classical Thermodynamics*) correlates a state of thermodynamic equilibrium with another state of thermodynamic equilibrium, however reached. It states that if the energy is varied this is due to flows of heat and work; that if the entropy has varied this is due to flows or productions of entropy. But it does not give information on what happens during the transformation if it does not go, instant by instant, through states of thermodynamic equilibrium (which are states in which flows and productions are identically zero).

In order to give meaning to flows and productions, and to derive operational engineering expressions from them, we build a new building (which includes the previous one of *classical Thermodynamics*), still on a postulate basis. First of all, let us formulate a new postulate that allows us to define internal energy, entropy and other thermodynamic quantities even in states of the system that are not in thermodynamic equilibrium.

[1.7] *At any point P of a system, called I a suitably small neighborhood of P, the subsystem identified by I is in a state of thermodynamic equilibrium.* ■

The main limitation of this postulate is that it implies that the history of the system, *i.e.* the processes previously undergone, do not influence the state of the system. In fact, it says that the internal energy in a point depends only on the values that the thermodynamic parameters have in that point. In nature, however, there is a vast phenomenology to which this postulate cannot be applied. For example, in *hysteresis processes* and *heat treatments* the material properties depend on the previous history.

The postulate [1.7] allows to define the quantities of *classical Thermodynamics* even in states that are not in thermodynamic equilibrium. We now need tools that allow us to evaluate the flows and productions of these quantities. We therefore postulate now (*Curie principle*)

[1.8] *If the system is isotropic each flow depends on all and only the forces generalized by the same tensor order* [1.9] *of the flow.* ■

Considering that in the following we will always deal with isotropic systems, and that we call *generalized force* [1.10] any cause that can cause flows, the postulate [1.8] tells us that a flow which is a scalar [resp. vector] [resp. tensor] depends on all generalized forces which are scalars [resp. vectors] [resp. tensors] and only from those.

For example, in combustion, chemical affinity is a generalized force of the scalar type. It pushes the

---

[1.9] A quantity (associated to a point) is said to be *scalar* (or *tensor of order zero*) if it is identified by a real number; *vectorial* (or *tensor of order* 1) if it is identified by an ordered triad of real numbers; *tensor* (or *tensor of order* 2, or simply *tensor*) if it is identified by an ordered sextuple of real numbers (that is, by two vectors). It should be noted that a direction is associated with a vector; two directions are associated with a tensor (or, if you prefer, a plane and a direction).

[1.10] This is because the *work flow* is caused by a cause that we call a *force*.

reaction one way or the other. However, being a scalar, it cannot influence (if the medium is isotropic) neither the energy flows nor the mass flow (which are vector quantities).

Normally a generalized force causes one flow directly and other flows indirectly. For example, the generalized force $\dfrac{\partial}{\partial r}\dfrac{1}{T}$ is the direct cause of the flow of energy in the form of heat and is an indirect cause of the flow of mass.

What is the functional dependence of the flows on the causes remains to be clarified. We recall that (under suitable hypotheses) a real function of a real variable can be developed in a *McLaurin* series

$$(1.10) \qquad f(c) = f(0) + f'(0)\, c + f''(0)\, \frac{c^2}{2} + \cdots .$$

We express for the effect (*i.e.* for the flow) this functional dependence on the cause (*i.e.* on the generalized force). We choose the cause so that when

the cause is zero the effect is zero. Therefore $f(0) = 0$.
We limit the study to the linear thermodynamics of
irreversible processes (so that in (1.10) we neglect the
terms of order greater than 1).

We have come to the conclusion that the effect is
proportional to the cause and, most importantly, the
coefficient (of proportionality) is independent of the
cause. Therefore $f'(0)$ is a purely thermodynamic
quantity, *i.e.* it is a function of state. For example, in
Fourier's law of heat transfer

$$(1.11) \qquad\qquad Q = k\,\frac{\partial T}{\partial}r$$

the constant k is a state function of the medium (of
course this law is valid only if there is the possibility of
mass flows).

Thus, in an isotropic system the fluxes depend on
all the generalized forces of the same tensor order (direct
and indirect). If the phenomenology is linear then the
flows are proportional to all the causes that can cause

them. Furthermore, the proportionality coefficients are state functions, *i.e.* they depend only on the state of the system.

Let us consider a system consisting of a mixture of two gases. There is energy flow in the form of heat Q and mass flow Φ_{M_1} of the first gas. The relative direct generalized forces are $\frac{\partial}{\partial r}\left(\frac{1}{T}\right)$ and $\frac{\partial}{\partial r}\left(\frac{\mu}{T}\right)$. According to *Curie*'s principle, since these directed generalized forces are vectors and both fluxes are vectors, each flux must depend on both generalized forces. Then, in the field of linear thermodynamics we will write

$$Q = L_{qq}\,\frac{\partial}{\partial r}\left(\frac{1}{T}\right) + L_{qm}\,\frac{\partial}{\partial r}\left(\frac{\mu}{T}\right)$$

$$Q = L_{mq}\,\frac{\partial}{\partial r}\left(\frac{1}{T}\right) + L_{mm}\,\frac{\partial}{\partial r}\left(\frac{\mu}{T}\right).$$

The direct proportionality coefficients L_{qq}, L_{mm} and the cross ones L_{qm}, L_{mq} are functions only of the state of the system.

Aldo Maceri

The following *Onsager* theorem holds. It has general validity, even for systems with n degrees of freedom.

[1.9] *In the absence of magnetic fields the crossed coefficients are equal:*

$$L_{qm} = L_{mq} \; . \; \blacksquare$$

These conclusions are of the utmost practical importance. For example, it is thanks to them that it is possible to build refrigerators based on the thermoelectric effect. In them the exchange of energy in the form of heat is obtained by means of an electric potential difference.

It is shown

$$(1.12) \qquad \delta_S = \sum_{i=1}^{n} \Phi_i \cdot F_i \; .$$

In (1.12) δ_S is the local entropy production; $\Phi_1, ..., \Phi_n$

are all the fluxes involved; $F_1, ..., F_n$ are all the generalized forces involved. In (1.12) the products are scalar if flows and forces are vectors; scalar double products if flows and forces are tensors. We observe that in (1.12) the flow of energy in the form of work does not appear because it is not accompanied by the production of entropy.

We postulated that $\delta_S \geq 0$. This results in conditions on the direct and cross coefficients L_{ij} . For example, in the already examined case in which they are L_{qq} , L_{mm} , L_{qm} , L_{mq} it results:

$$L_{qq} \geq 0$$
$$L_{mm} \geq 0$$
$$L_{qq} L_{mm} - L_{mq}^2 \geq 0 \ .$$

Chapter 2

Fluid Mechanics

2.1   The equations of *Fluid Dynamics*

We begin the study of *Fluid Dynamics* with the following premises. Let $\boldsymbol{u}$ and $\boldsymbol{v}$ be two vectors of $\Re^3$ with components (u_x, u_y, u_z), (v_x, v_y, v_z). It is called the *tensor product of* $\boldsymbol{u}$ by $\boldsymbol{v}$ (and is denoted by $\boldsymbol{u} \cdot \boldsymbol{v}$) the tensor of $\Re^3$

$$\begin{bmatrix} u_x v_x & u_x v_y & u_x v_z \\ u_y v_x & u_y v_y & u_y v_z \\ u_z v_x & u_z v_y & u_z v_z \end{bmatrix}.$$

Let us consider a *tensor* $\boldsymbol{\tau}$ of $\Re^3$

$$\begin{bmatrix} \tau_{xx} & \tau_{xy} & \tau_{xz} \\ \tau_{yx} & \tau_{yy} & \tau_{yz} \\ \tau_{zx} & \tau_{zy} & \tau_{zz} \end{bmatrix}$$

and a *vector* $\boldsymbol{n}$ of $\mathfrak{R}^3$

$$\boldsymbol{n} = \begin{bmatrix} n_x \\ n_y \\ n_z \end{bmatrix}.$$

We call the *scalar product* of $\boldsymbol{\tau}$ by $\boldsymbol{n}$ (and denote by the symbol $\boldsymbol{\tau} \times \boldsymbol{n}$) the vector of $\mathfrak{R}^3$

$$\boldsymbol{p} = \begin{bmatrix} \tau_{xx} & \tau_{xy} & \tau_{xz} \\ \tau_{yx} & \tau_{yy} & \tau_{yz} \\ \tau_{zx} & \tau_{zy} & \tau_{zz} \end{bmatrix} \cdot \begin{bmatrix} n_x \\ n_y \\ n_z \end{bmatrix}$$

$$= \begin{bmatrix} \tau_{xx}n_x & \tau_{xy}n_y & \tau_{xz}n_z \\ \tau_{yx}n_x & \tau_{yy}n_y & \tau_{yz}n_z \\ \tau_{zx}n_x & \tau_{zy}n_y & \tau_{zz}n_z \end{bmatrix}.$$

Calling $\boldsymbol{x}, \boldsymbol{y}, \boldsymbol{z}$ the *unit vectors* of the coordinate axes (so that $\boldsymbol{p} = p_x\boldsymbol{x} + p_y\boldsymbol{y} + p_z\boldsymbol{z}$), we call the *flux of the tensor* $\boldsymbol{\tau}$ *through* Σ the vector

$$\int_\Sigma \boldsymbol{\tau} \times \boldsymbol{n} \; d\sigma = \int_\Sigma \boldsymbol{p} \; d\sigma =$$

$$= \left(\int_\Sigma p_x \, dx \right) x + \left(\int_\Sigma p_y \, dy \right) y + \left(\int_\Sigma p_z \, dz \right) z \ .$$

After that, we highlight that any fluid dynamics problem has two types of unknowns: the state unknowns (which characterize the thermodynamic state of the system) and the kinetic unknown (which is the velocity vector). Obviously, these unknowns are functions of point and time.

It goes without saying that it is necessary to formulate as many scalar equations as there are the thermodynamic unknowns of state plus a vector equation in which the velocity vector intervenes. To formulate these equations, we impose the fact that the evolution of the system takes place by satisfying the conservation or balance equations.

If the system consists of only one phase, we impose conservation of mass. If instead it is a mixture of different masses, we also impose the balance of the single masses (or the conservation of the single masses, if the possibility of chemical reactions is excluded). Another scalar equation is that of the conservation of total energy. A vector equation (which breaks down into as many scalar equations as there are components of the vector) is the momentum balance. The production of momentum is due to mass forces (such as the force of gravity). In the absence of such forces, the conservation of momentum will prevail.

Let us therefore explain the balance of the total mass M. Calling ρ the mass density, $\boldsymbol{v}$ the velocity and $\boldsymbol{n}$ the unit vector of a straight-line n, we have

$$G = M \ , \ g = \rho \ , \ \varphi_M = \rho \ \boldsymbol{v} \times \boldsymbol{n} \ , \ \delta_M = 0 \ .$$

Since the local production δ_M is zero, the balance becomes a conservation and is written

(2.1) $$\frac{\partial}{\partial t} \int_V \rho \ dV + \int_\Sigma \rho \ \boldsymbol{v} \times \boldsymbol{n} \ d\sigma = 0 \ .$$

Reasoning in the same way, applying the mass balance to an elementary parallelepiped (with faces parallel to the coordinate planes) and taking into account the divergence theorem, we obtain the differential equation in $V \times [t_i, +\infty[$

$$\frac{\partial \rho}{\partial t} + div(\rho \boldsymbol{v}) = 0 \ .$$

We explain the momentum balance. The momentum density is $g = \rho \boldsymbol{v}$. From the weighted mean theorem (of the *Integration Theory*) and from the definition of barycenter it easily follows that the total momentum is $G = M\boldsymbol{v_0}$ (where $\boldsymbol{v_0}$ is the speed of the barycenter of the system).

The flux of momentum is the flux of a tensor. In fact, just as the flux of a scalar quantity G is the flux of a vector $g\boldsymbol{v}$, so the flux of a vector quantity (which is the momentum) is the flux of a tensor.

We split the momentum flux $\Phi_{\rho v}$ into an aliquot

2.1

$$\int_{\Sigma} (\rho \, \boldsymbol{v} \cdot \boldsymbol{v}) \times n \; d\sigma$$

which we call *convective momentum flux* and in an aliquot which we call *diffusive momentum flux*. This diffusive momentum flux is also the flux of a tensor, which we denote by the symbol $\boldsymbol{\tau}$:

$$\int_{\Sigma} \boldsymbol{\tau} \times \boldsymbol{n} \; d\sigma$$

The tensor $\boldsymbol{\tau}$ has the dimensions of a force per unit area and is just the stress tensor. Precisely, on the macroscopic level the result of the molecular agitations, with respect to the transport of the momentum, is manifested with a system of surface efforts.

If there is no production of momentum, balancing the momentum of momentum we get the symmetry of

---

[2.1] See what is mentioned in the introduction.

the $\boldsymbol{\tau}$.

Finally, the production of momentum is due to the specific *mass* (also called *volume*) *forces* $\boldsymbol{f}$ but not to the *surface* (also called *contact*) ones:

$$\Delta_{\rho v} = \int_V f \, dV \ .$$

Ultimately, the momentum balance equation is written

$$(2.2) \qquad \frac{\partial}{\partial t} \int_V \rho v \, dV + \int_\Sigma (\rho \, \boldsymbol{v} \cdot \boldsymbol{v} + \boldsymbol{\tau}) \times \boldsymbol{n} \, d\sigma$$

$$= \int_V f \, dV \ .$$

We explain the conservation of *total energy*. A system has many forms of energy. The *total energy* of the system is conserved. A first contribution to the total energy is the *internal energy U* of the system. A second contribution is *kinetic energy*

$$\int_V \frac{1}{2}\,\rho\,v^2\,dV \ .$$

In the absence of gravitational and electromagnetic fields, the sum of these two contributions constitutes the total energy of the system. It is important to highlight that (since kinetic energy can be converted into internal energy and vice versa) each of these two contributions may not be conserved. The difficulty of formulating the equation of conservation of total energy lies precisely in recognizing which are the significant contributions that intervene in a given phenomenon. For example, in a nuclear reaction it is necessary to take into account the change in energy resulting from the change in mass.

Let us consider a system in which the total energy is the sum of the internal energy and the kinetic energy:

$$G = U + \frac{1}{2}Mv^2.$$

The density of G is

$$g = u + \frac{1}{2}\rho v^2 ,$$

where u is the internal energy per unit mass.

The *convective flux* of the total energy is

$$gv = \rho \left(u + \frac{v^2}{2} \right) v .$$

We denote by J_t the *diffusive flux* of the total energy. It is the sum of three addends: the diffusive flow of energy associated with any mass flows, the diffusive flow of energy in the form of heat and the diffusive flow of energy in the form of work.

The diffusive flux of energy in the form of work is the scalar product of the surface stress tensor and the velocity vector with which the point of application of the surface stress moves.

The diffusive flux of energy in the form of heat is what we previously denoted by the symbol Q .

The diffusive flux of energy associated with the mass flux is given by the mass flux multiplied by the energy per unit mass related to the chemical potential of the substance.

The diffusive flow of energy in the form of work $\tau \times v$ can be divided into a reversible part (which is not associated with the production of entropy) and an irreversible part.

The diffusive flow of energy in the form of heat is always irreversible (that is, it is always accompanied by a production of entropy).

It is clear that to split the diffusive work flow $\tau \times v$ into a reversible part and an irreversible part it is necessary to split the stress tensor [2.2]

$$\tau = \begin{bmatrix} \sigma_z & \tau_{xy} & \tau_{xz} \\ \tau_{yx} & \sigma_y & \tau_{yz} \\ \tau_{zx} & \tau_{zy} & \sigma_z \end{bmatrix}$$

[2.2] As known, the component τ_{xx} [resp. τ_{yy}] [resp. τ_{zz}] of τ can also be denoted by the symbol σ_x [resp. σ_y] [resp. σ_z].

into a reversible part and an irreversible part. We mean by reversible [resp. irreversible] stress that part of the stress which gives rise to reversible [resp. irreversible] work.

To recognize the reversible part of $\boldsymbol{\tau}$ we refer to *classical Thermodynamics*, which by definition is the *Thermodynamics of reversible phenomena*. In it the effort taken into consideration is the *equilibrium pressure*.

Therefore, we decompose the stress tensor $\boldsymbol{\tau}$ into the sum of a hydrostatic tensor $\boldsymbol{p}$ and a remaining aliquot $\boldsymbol{\tau_d}$. Thereby

$$\boldsymbol{\tau} \times \boldsymbol{v} = \boldsymbol{p} \times \boldsymbol{v} + \boldsymbol{\tau_d} \times \boldsymbol{v} = \boldsymbol{pv} + \boldsymbol{\tau_d} \times \boldsymbol{v} \ .$$

Only the *constitutive equations* remain to be added, which express the mechanical behavior of the medium.

To characterize the state of a fluid, it is sufficient to assign the volume. To characterize the state of a solid it is necessary to assign not the volume, but its

Fluid Dynamics

deformation state. Thus, in fluids the equation of state is given to express the dependence of p on the parameters representing the system. In solids there are relations (for example those of *Navier*) which express the dependence of the stress tensor on the strain tensor.

In any case, given the expression of τ_d , the system of equations which solves every problem of the continuum (both fluid and solid) will be closed, having as many equations as there are unknowns.

Ultimately, in the absence of mass diffusive fluxes and in the absence of potential energy deriving from mass forces, the equation of conservation of total energy is written

(2.3)
$$\frac{\partial}{\partial t} \int_V \rho \left(u + \frac{v^2}{2} \right) dV +$$

$$+ \int_\Sigma \left[\left(u + \frac{v^2}{2} \right) \rho v + p v + \tau_d \times v + Q \right] \times n \, d\sigma = 0 .$$

Each term of the energy equation has the dimensions of

a power, *i.e.* of an energy per unit of time.

In any fluid dynamics (or solid mechanics) problem, it will be the *Engineer*'s job to simplify the problem by discarding those terms he deems appropriate. The choice of the terms to be discarded can be carried out with the *Theory of characteristic numbers*, which is mentioned here. As we will see, these numbers give a relative measure of the quantities involved in the problem, so that they allow us to identify the terms of the equations that can be neglected (within the approximation with which the problem is treated).

2.2   The characteristic numbers

We remove the dimensions to continuum equations. We choose the units of measure, a reference density ρ_r (so that $\rho = \rho^* \rho_r$, where ρ^* is a dimensionless number) and a reference speed v_r (so that $v = v^* v_r$, where v^* is a dimensionless number).

The v_r must be chosen in relation to the

phenomenon being studied. For example, if relativistic phenomena are studied, the v_r will be the speed of light because in those phenomena we are interested in what relationship the speed of the system has with that of light.

The engineer's task lies precisely in the appropriate choice of the reference quantity, that is, in grasping the physical meaning of the phenomenon being considered. For example, if the phenomenon of sound propagation is being studied, it would be wrong to choose the speed of light as the reference speed because all the numbers would be extremely small and therefore negligible and the phenomenon could not be studied. It is clear that in this second problem the reference speed is that of sound a , which is a thermodynamic quantity. In the air and in standard conditions (of temperature and pressure), it results $a = 340 \; \frac{m}{s}$. The same applies to time: in the study of a phenomenon involving sinusoidal oscillations, we will choose the period as the reference time t_r, and we will write $t = t^* t_r$, with t^* dimensionless number. Furthermore, if we study the

Aldo Maceri

heat transmission in a cylinder, the reference surface is the lateral one (proportional to the diameter). If instead we study the mass flow through the cylinder, the reference surface is the base (proportional to the square of the diameter).

Let us remove the dimensions to continuity equation. Turns out

$$\frac{\rho_r V_r}{t_r} \frac{\partial}{\partial t^*} \int_{V^*} \rho^* \, dV^* + \rho_r V_r \Sigma_r \int_{\Sigma^*} \boldsymbol{v}^* \times \boldsymbol{n} \, d\sigma = 0$$

and from here

(2.4)
$$\frac{\partial}{\partial t^*} \int_{V^*} \rho^* \, dV^*$$

$$+ \left(\frac{v_r \Sigma_r t_r}{V_r}\right) \int_{\Sigma^*} \rho^* \boldsymbol{v}^* \times \boldsymbol{n} \, d\sigma = 0 \ .$$

In (2.4) the first addendum is a dimensionless number and the two factors constituting the second addendum are also dimensionless. The first of these factors is called the *Strouhal* number and is denoted by the symbol S_{tr}

$$S_{tr} = \frac{v_r \Sigma_r t_r}{V_r} .$$

It measures the relative importance of the contribution of flow versus the contribution of unsteadiness in the equation of conservation of mass. If it is very small with respect to unity, the contribution of the flux is negligible with respect to the contribution of the unsteadiness; conversely, if it is very large, the unsteady term will be negligible with respect to the term relating to the flow. If S_{tr} has order of magnitude 1, the unsteady term and that relating to the flow are of the same order of magnitude, therefore both must be taken into account.

All characteristic numbers (being dimensionless) can always be interpreted as the ratio of two quantities having the same dimensions. Thus, it is possible to give various interpretations to these numbers.

The most immediate interpretation of the *Strouhal* number, since we are studying the importance of convection with respect to unsteadiness, is that of the ratio between two times. We will call the *reference macroscopic time* (or *convective time*) the term $\dfrac{V_r}{v_r \Sigma_r}$ (which has exactly the dimensions of a time). It is the characteristic time of convection and measures the average permanence of the particle in the control volume V_r . The time t_r instead is characteristic of the unstationarity and measures the speed of the unsteadiness. If S_{tr} is very small the convection is negligible with respect to the unstationarity and the regime is said to be *unsteady*.

Remark 2.1 We note that despite having obtained the *Strouhal* number from the mass conservation equation, mass does not appear in it. In reality, if we had written the conservation equation of any other quantity G , we would have arrived (as it is immediate to verify) at the same expression of S_{tr}. This means that the relative importance of convection with respect to unsteadiness is always the same whatever the quantity that flows. ∎

We remove the dimensions to momentum balance equation, given by (2.2), which we rewrite

$$\frac{\partial}{\partial t} \int_V \rho \boldsymbol{v} \, dV + \int_\Sigma (\rho \, \boldsymbol{v} \cdot \boldsymbol{v} + \boldsymbol{\tau}) \times \boldsymbol{n} \, d\sigma = \int_V f \, dV \, .$$

We get

$$(2.5) \qquad \frac{\rho_r v_r V_r}{t_r} \frac{\partial}{\partial t^*} \int_{V^*} \rho^* v^* \, dV^*$$

$$+ \rho_r v_r^2 \Sigma_r \int_{\Sigma^*} \rho^* v^* \cdot v^* \times \boldsymbol{n} \, d\sigma$$

$$+ p_r \Sigma_r \int_{\Sigma^*} \boldsymbol{p} \times \boldsymbol{n} \, d\sigma + \tau_{dr} \Sigma_r \int_{\Sigma^*} \boldsymbol{\tau_d} \times \boldsymbol{n} \, d\sigma$$

$$= f_r V_r \int_{V^*} f^* \, dV^* \, .$$

As it is easy to verify, in (2.5) the coefficients of the dimensionless factors have the dimensions of a momentum per unit of time, *i.e.* of a force. Dividing the first and second sides of (2.5) by $\rho_r v_r^2 \Sigma_r$ we obtain

$$(2.6) \quad \frac{1}{S_{tr}} \frac{\partial}{\partial t^*} \int_{V^*} \rho^* v^* \, dV^* + \int_{\Sigma^*} \rho^* v^* \cdot v^* \times n \, d\sigma$$

$$+ \frac{p_r}{\rho_r v_r^2} \int_{\Sigma^*} p \times n \, d\sigma + \frac{\tau_{dr}}{\rho_r v_r^2} \int_{\Sigma^*} \tau_d \times n \, d\sigma$$

$$= \frac{f_r V_r}{\rho_r v_r^2 \Sigma_r} \int_{V^*} f^* \, dV^* \ .$$

We observe that in the three new characteristic numbers the reference surface appears only in the third. In fact, it measures the relative importance between the volume contribution and the surface one.

About the characteristic number

$$\frac{p_r}{\rho_r v_r^2}$$

two cases must be distinguished. In the case of an incompressible fluid, the hydrostatic pressure must be assumed as the reference pressure. However, if the phenomenon of *cavitation* is being studied, p_r is the surface vapor tension. In this case $\frac{p_r}{\rho_r v_r^2}$ is called the *Weber* number.

Remark 2.2 It is important to note that the *Weber* number can be varied by means of chemical additives which vary the value of the surface vapor tension. The *Weber* number is important for the study of dual-phase mass flows, in which the liquid phase and the vapor phase coexist. In them the vapor pressure is important for establishing the equilibrium conditions of a vapor bubble. ∎

In the case of compressible fluids, the reference pressure is the thermodynamic pressure $\frac{\partial U}{\partial V}$. It is the macroscopic result of the molecular agitations for which it must be proportional to the average speed of the

Aldo Maceri

molecules, which coincides with that of sound a . If the compressible fluid is a *perfect* gas [2.3] results

$$p = \rho RT \quad , \quad a^2 = \gamma RT$$

where $\gamma = \dfrac{c_p}{c_V}$, $R = c_p - c_V$. Then, $p = \dfrac{a^2 \rho}{\gamma}$ and the characteristic number is written

$$\frac{p_r}{\rho_r v_r^2} = \frac{a^2}{\gamma v_r^2} = \frac{1}{\gamma M^2} .$$

Thus, in compressible fluids the relative importance of non-dissipative momentum flow with respect to convective flow depends on γ and on the (dimensionless) characteristic number

$$M = \frac{v_r}{a}$$

which we call number of *Mach*.

[2.3] A perfect gas without viscosity is called a *more than perfect* gas.

We give a *kinetic* interpretation of the *Mach* number. M is the ratio between the macroscopic speed of the medium and the speed of sound within the medium itself. The speed of sound is the speed with which small disturbances (such as small pressure variations) propagate in the medium. We can interpret v_r as the speed with which small disturbances are created. If a body moves in a continuous medium with speed v_r, it creates small disturbances with speed v_r (which propagate with speed a) in the medium. If the body moves at *Mach* < 1 (*subsonic flow*), it generates a pressure wave that is perceived by the medium before the arrival of the body. Thus, the vehicle knows of the arrival of the body before it has arrived and in a certain way prepares to welcome it, creating free space for its passage. If instead the body moves at *Mach* >1 (*supersonic flow*) the body will hit the medium before the pressure waves it produces. Thus, the environment, having no way of noticing the presence of the body before its arrival, will be forced to open abruptly to allow the body to pass.

The same considerations can be made if it is not the body that moves in the environment, but the fluid current that invests the body.

Given the different behavior of the subsonic flow and the supersonic flow, a singular behavior in correspondence of $M = 1$ (*transonic flow*) is to be expected (and this is exactly what happens).

We give a *dynamic* interpretation of the *Mach* number. We have seen that for compressible fluids

$$\frac{p_r}{\rho_r v_r^2} = \frac{1}{\gamma M^2} \; .$$

Since γ always has order of magnitude 1, M gives a measure of the importance of the resultant of the pressure forces with respect to that of the inertia forces.

We give an *elastic* interpretation of the *Mach* number (but more than an interpretation of the *Mach* number, this is an interpretation of the speed of sound).

Small disturbances are propagated through a succession of compressions and expansions of the medium, so that the speed of sound is greater the more compressible the medium is. This can also be seen from the relation (with constant entropy)

$$a^2 = \frac{\partial p}{\partial \rho} \ .$$

So, if $M \approx 0$ either the body is practically stationary or the motion takes place in a highly incompressible medium. Better said, the medium behaves as if it were incompressible. If the speed with which the perturbations are produced is very small it is $M \approx 0$ and the air, which is a highly compressible fluid, behaves like an incompressible fluid. Conversely, water, which is a fluid with little compressibility, will behave like a highly compressible fluid when the speed with which the perturbations are produced is very high (*water hammer*).

We give an *energetic* interpretation of the *Mach* number. So long as

$$M^2 = \frac{\rho_r v_r^2 v_r}{\gamma p_r v_r} \, ,$$

energetically M measures the relative importance between the convective flow of kinetic energy and the reversible work flow.

Let us now consider, in (2.6), the characteristic number

$$\frac{\tau_{dr}}{\rho_r v_r^2}$$

which represents the importance of the dissipative diffusive momentum flux compared to the convective flux.

To get an idea of the criterion according to which τ_{dr} is to be chosen, we must first understand what τ_d is, *i.e.* the dissipative part of the stress tensor $\boldsymbol{\tau} = \boldsymbol{p} + \boldsymbol{\tau_d}$ that we introduced when we performed the momentum balance. We said that $\boldsymbol{\tau}$ is the diffusive flux of momentum and when we performed the conservation of energy we found this same surface stress $\boldsymbol{\tau_d}$ (in the diffusive flux of energy in the form of work) and we distinguished a *non-dissipative* or *isotropic part* $\boldsymbol{p}$ made up of the *thermodynamic pressure* and a *dissipative part* $\boldsymbol{\tau_d}$. Being a dissipative local momentum flow, $\boldsymbol{\tau_d}$ is of the same nature as $\boldsymbol{Q}$ and the entropy flow, *i.e.* it is a quantity that needs the *Thermodynamics of irreversible processes* to be explained.

Proceeding along this path, we must identify the direct generalized force that produces this dissipative flow and (invoking the *Curie* principle) say that since this flow is a tensor quantity, it must depend (in an isotropic system) on all and only the generalized forces of the type tensor, linearly in the hypothesis of linear phenomenology.

Aldo Maceri

In this particular case, however, we prefer to search for the expression of $\boldsymbol{\tau_d}$ (*i.e.* the *constitutive equation* of the medium) by proceeding in another way.

Preliminarily we assume that the tensor $\boldsymbol{\tau_d}$ is symmetric, so that

$$\tau_{dxy} = \tau_{dyx} \ , \quad \tau_{dxz} = \tau_{dzx} \ , \quad \tau_{dyz} = \tau_{dzy} \ .$$

We now relate the stress $\boldsymbol{\tau}$ to the strain of the fluid. We note that while in the case of solids $\boldsymbol{\tau}$ depends (linearly in the *Navier* model) on the deformation, in that of fluids the surface stress does not already depend on the deformation, but on the speed of deformation. In fact, if we slowly immerse a hand in a liquid, we do not feel any stress on the hand. If, on the other hand, the hand is immersed rapidly, we feel a stress that is all the more intense the greater the immersion speed (that is, the greater the fluid deformation speed).

With this reasoning we have identified the direct generalized force of $\boldsymbol{\tau_d}$. In the hypothesis of linearity, we assume for fluids (in the hypothesis of isotropy)

$$\sigma_{dx} = 2\mu \frac{\partial \varepsilon_x}{\partial t} + \mu_2 \frac{\partial}{\partial t}\left(\varepsilon_x + \varepsilon_y + \varepsilon_z\right)$$

$$\sigma_{dy} = 2\mu \frac{\partial \varepsilon_y}{\partial t} + \mu_2 \frac{\partial}{\partial t}\left(\varepsilon_x + \varepsilon_y + \varepsilon_z\right)$$

$$\sigma_{dz} = 2\mu \frac{\partial \varepsilon_z}{\partial t} + \mu_2 \frac{\partial}{\partial t}\left(\varepsilon_x + \varepsilon_y + \varepsilon_z\right)$$

$$\tau_{dxy} = \mu \frac{\partial \gamma_{xy}}{\partial t}$$

$$\tau_{dxz} = \mu \frac{\partial \gamma_{x}z}{\partial t}$$

$$\tau_{dyz} = \mu \frac{\partial \gamma_{yz}}{\partial t} \ .$$

Therefore, given u_x , u_y , u_z the components of the displacement of the fluid particle and v_x , v_y , v_z the components of its velocity v (so that $v_x = \frac{\partial u_x}{\partial t}$, $v_y = \frac{\partial u_y}{\partial t}$, $v_z = \frac{\partial u_z}{\partial t}$), resulting in *Deformation Analysis* (in a continuous medium and for small displacements)

$$\varepsilon_x = \frac{\partial u_x}{\partial x}$$

$$\varepsilon_y = \frac{\partial u_y}{\partial y}$$

$$\varepsilon_z = \frac{\partial u_z}{\partial z}$$

$$\gamma_{xy} = \frac{\partial u_x}{\partial y} + \frac{\partial u_y}{\partial x}$$

$$\gamma_{xz} = \frac{\partial u_x}{\partial z} + \frac{\partial u_z}{\partial x}$$

$$\gamma_{yz} = \frac{\partial u_y}{\partial z} + \frac{\partial u_z}{\partial y} \ ,$$

we have

$$(2.7) \qquad \sigma_{dx} = 2\mu \frac{\partial v_x}{\partial x} + \mu_2 \, div(\boldsymbol{v})$$

$$\sigma_{dy} = 2\mu \frac{\partial v_y}{\partial y} + \mu_2 \, div(\boldsymbol{v})$$

$$\sigma_{dz} = 2\mu \frac{\partial v_z}{\partial z} + \mu_2 \, div(\boldsymbol{v})$$

$$\tau_{xy} = \mu \left(\frac{\partial v_x}{\partial y} + \frac{\partial v_y}{\partial x} \right)$$

$$\tau_{xz} = \mu \left(\frac{\partial v_x}{\partial z} + \frac{\partial v_z}{\partial x} \right)$$

$$\tau_{yz} = \mu \left(\frac{\partial v_y}{\partial z} + \frac{\partial v_z}{\partial y} \right) \ .$$

The coefficients μ and μ_2 are called *first* and *second viscosity coefficients* and are state functions. Therefore, they do not depend on the strain rate, but only on the medium and the parameters that identify the state of the system.

The dependence of μ on temperature is different for liquids and for gases. Precisely for liquids μ decreases as the temperature increases while the opposite occurs for gases. Furthermore, the viscosity of gases is always considerably lower than that of liquids.

Remark 2.3 This different behavior is explained on a molecular level by the fact that viscosity is due to two factors: molecular attraction and thermal agitation. In liquids, viscosity is essentially due to the first factor, whereby increasing the temperature decreases the molecular attraction and with it decreases the viscosity. In gases, on the other hand, the viscosity is essentially due to thermal agitation, so as the temperature increases, the thermal agitation increases and with them the viscosity increases. ■

Remark   2.4   We observe that, resulting

$$div(v) = \frac{\partial}{\partial t}\left(\varepsilon_x + \varepsilon_y + \varepsilon_z\right) ,$$

$div(v)$ is the change in volume per unit volume and per unit time, so we can also call it *volume production*. ■

Remark   2.5   We also observe that, since

$$p_d = \frac{\sigma_{dx} + \sigma_{dy} + \sigma_{dz}}{3} = \left(\frac{2}{3}\mu + \mu_2\right) div(v) ,$$

p_d is *invariant* with reference and depends only on $div(v)$. ■

In most gases

$$\frac{2}{3}\mu + \mu_2 \cong 0$$

so that $p_d \cong 0$ and consequently for the stress tensor $\boldsymbol{\tau} = \boldsymbol{p} + \boldsymbol{\tau_d}$ we obtain

$$\frac{\sigma_x + \sigma_y + \sigma_z}{3} = p$$

where p is the thermodynamic pressure.

In conclusion, in fluids the direct generalized force of $\boldsymbol{\tau_d}$ is the tensor $^{2.4}$ $\frac{\partial v}{\partial r}$ (while in solids it is the tensor $\boldsymbol{\varepsilon}$).

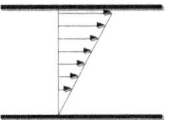

Fig. 2.1

$^{2.4}$ We denote $\dfrac{\partial v}{\partial r} = \begin{bmatrix} \dfrac{\partial v_x}{\partial x} & \dfrac{\partial v_x}{\partial y} & \dfrac{\partial v_x}{\partial z} \\[2mm] \dfrac{\partial v_y}{\partial x} & \dfrac{\partial v_y}{\partial y} & \dfrac{\partial v_y}{\partial z} \\[2mm] \dfrac{\partial v_z}{\partial x} & \dfrac{\partial v_z}{\partial y} & \dfrac{\partial v_z}{\partial z} \end{bmatrix}$.

Aldo Maceri

Equations (2.7) are a generalization of *Newton's* law

$$F = \mu \, A \, \frac{dv}{dn}$$

on the friction offered by a fluid interposed between two parallel flat surfaces (of area A) in relative motion (fig. 2.1). The fluids which obey (2.7) are therefore called *Newtonian*. And yet there are many non-*Newtonian* fluids (such as heavy lubricating oils).

Remark   2.6   In *Mechanics of Solids*, in the elastic range all solids are *Hookian* (in the sense that they obey *Navier*'s relations). ■

After these premises, we pass to the choice of τ_{dr} in the characteristic number

$$\frac{\tau_{dr}}{\rho_r v_r^2} \, .$$

Suppose (which is not always the case) that the fluid is

Fluid Dynamics

Newtonian. To choose τ_{dr} , resulting for (2.7)

$$\tau_{dr} = \mu_r \frac{v_r}{L_r} \; ,$$

a reference viscosity coefficient, a reference speed and a reference length must be chosen. The μ_r is a state function and can be chosen in the initial state. With this the characteristic number of (2.6)

$$\frac{\tau_{dr}}{\rho_r v_r^2}$$

turns to

$$\frac{\mu_r}{\rho_r v_r L_r} \; .$$

Its reciprocal is a dimensionless characteristic number, it is called the *Reynolds* number and is denoted by the symbol

$$R_e = \frac{\rho_r v_r L_r}{\mu_r} \; .$$

It is also customary to introduce the *kinematic viscosity*

$$\nu_r = \frac{\mu_r}{\rho_r}$$

and express with it R_e , as follows

$$R_e = \frac{v_r L_r}{\nu_r} \ .$$

Unlike the *Mach* number, the *Reynolds* number contains a length (which takes into account the geometry of the system). For a sphere hit by a fluid current, L_r will be the diameter of the sphere. For an array of turbine blades, L_r will be either the blade chord or the blade pitch (indifferently, since they are of the same order of magnitude). For a nozzle, L_r will be the throat diameter.

Remark 2.7 Normally the choice of the reference length (which must be consistent with the phenomenon being studied) is very delicate. ■

Fluid Dynamics

We give a *kinetic* interpretation of the *Reynolds* number. We write R_e in the form

$$R_e = \frac{v_r}{\left(\frac{v_r}{L_r}\right)}$$

so that R_e is the ratio of two speeds. The speed $\frac{v_r}{L_r}$ is the one with which the transversal waves propagate, *i.e.* the waves that propagate in the direction normal to the cause that produces them (for example the flexural vibrations of a tree). Transversal waves are due to the tangential components of surface stresses. Contrary to the speed of sound, the speed $\frac{v_r}{L_r}$ is a function of the thermodynamic state and geometry of the system.

We give a *dynamic* interpretation of the *Reynolds* number. We write R_e in the form

$$(2.8) \qquad R_e = \frac{\rho_r v_r^2}{\mu_r v_r / L_r} = \frac{\rho_r v_r^2}{\tau_{dr}} .$$

Aldo Maceri

Therefore R_e measures the importance of the *ordered (convective) momentum* flow compared to the *dissipative (diffusive) momentum* flow.

Note that M measures the relative importance between the *convective momentum* flux and the *non-dissipative diffusive momentum* flux, while R_e measures the relative importance between the *convective momentum* flux and the *dissipative momentum* flux of motion.

Note also that, in terms of forces, M measures the importance of the *resultant of the inertia forces* with respect to the *resultant of the non-dissipative normal stresses*, while R_e measures the importance of the *resultant of the inertia forces* with respect to the *resultant of the dissipative surface stresses*.

When R_e assumes very high values, it means that the importance of the resultant of the dissipative surface forces is negligible compared to the resultant of the inertia forces. In this case it will be possible to consider the fluid as if there were no dissipative surface stresses, with which the stress tensor $\boldsymbol{\tau}$ is reduced to the isotropic part $\boldsymbol{p}$ only.

The reverse happens for R_e close to 0. In this case (which is, for example, that of *lubrication*) the inertia forces are negligible with respect to the resultant of the dissipative surface stresses.

We give an *energetic* interpretation of the *Reynolds* number. From (2.8) it follows that R_e is the ratio between the kinetic energy (per unit of volume) and the dissipative work (per unit of volume).

Thus, at high *Reynolds* numbers the fluid is unable to exert its ability to do irreversible work.

We give an *entropic* interpretation of the *Reynolds* number. From the energetic interpretation of R_e it follows that the *Reynolds* number measures the effectiveness of viscosity in the production of entropy. This makes the concept of reversibility or irreversibility operational, because we have found (with R_e) the measure of the effectiveness of the cause that produces the entropy.

Let us now consider the last characteristic number appearing in equation (2.6) of the momentum balance

$$(2.9) \qquad\qquad \frac{f_r V_r}{\rho_r v_r^2 \Sigma_r} \; .$$

It represents the measure of the relative importance of the *momentum production* with respect to the *convective flow*, *i.e.* its value allows us to say whether the momentum production is negligible or not with respect to the convective flow.

In (2.9) it is necessary to choose f_r , which is a force per unit of volume (capable of producing momentum). The choice of f_r depends on the type of force acting on the system. For electrically neutral fluids (that is, that have no free charges) the electromagnetic forces obviously don't matter. In this case as f_r we choose the force due to the gravitational field

$$f_r = \rho_r g$$

where g is the acceleration due to gravity and ρ_r is the reference density. With this the characteristic number (2.9) is written

$$\frac{f_r V_r}{\rho_r v_r^2 \Sigma_r} = \frac{g V_r}{v_r^2 \Sigma_r} \ .$$

In the particular case in which $V_r = L_r^3$ and $\Sigma_r = L_r^2$ can be set, the characteristic number (2.9) is called *Froude number*, designated by F_r and results

$$F_r = \frac{g L_r}{v_r} \ .$$

Remark 2.8 The *Froude* number is of paramount importance in naval applications. ∎

By placing the *Froude* number in the form

$$F_r = \frac{g}{v_r/L_r} \; ,$$

it can be seen that it measures the relative importance between the forces of gravity and those of inertia.

We have already highlighted that it is one thing to have the ability to create a certain effect and quite another thing to have the possibility and the time to produce that effect. In fact, we have seen that all fluids have viscosity, however (depending on the value of R_e) they can behave as if they had no viscosity.

Similarly, for the *Froude* number it is assumed that all fluids are heavy (that is, they are affected by the effects of the acceleration of gravity). However, when F_r is very small it is as if the fluid were not heavy, *i.e.* the effects of acceleration due to gravity are negligible.

Remark 2.9 Consider for example the case of a fire hydrant. If the water flows with a small speed, v_r is small and therefore F_r is large and the water is affected by the effects of gravity. In fact (as soon as it comes out of the hydrant) the jet immediately converges downwards. Conversely, when the speed of the water jet is high, it remains horizontal for a long stretch because it is not affected by the effects of gravity. Only in the farthest points, due to the decrease in the speed of the jet, the effects of gravity return to predominate. ∎

It is important to note that the density does not appear in the *Froude* number, but g , which is independent of the medium.

We remove now the dimensions in the energy conservation equation (2.3), which we rewrite

$$\frac{\partial}{\partial t} \int_V \rho \left(u + \frac{v^2}{2} \right) dV + \int_\Sigma \left[\left(u + \frac{v^2}{2} \right) \rho v + p v + \tau_d \times v + Q \right] \times n \, d\sigma = 0 \ .$$

We get as a result

Aldo Maceri

$$(2.10) \qquad \frac{\rho_r u_r V_r}{t_r} \frac{\partial}{\partial t^*} \int_{V^*} \rho^* u^* \, dV^*$$

$$+ \frac{\rho_r v_r^2 V_r}{t_r} \int_{V^*} \rho^* \frac{(v^*)^2}{2} \, dV^*$$

$$+ \rho_r u_r v_r \Sigma_r \int_{\Sigma^*} u^* \rho^* \boldsymbol{v}^* \times \boldsymbol{n} \, d\sigma$$

$$+ \rho_r v_r^3 \Sigma_r \int_{\Sigma^*} \frac{1}{2} \overset{*}{\rho} (v^*)^2 \overset{*}{\boldsymbol{v}} \times \boldsymbol{n} \, d\sigma$$

$$+ p_r v_r \Sigma_r \int_{\Sigma^*} p^* \boldsymbol{v}^* \times \boldsymbol{n} \, d\sigma$$

$$+ \tau_{dr} v_r \Sigma_r \int_{\Sigma^*} (\boldsymbol{\tau_d} \times \boldsymbol{v}) \times \boldsymbol{n} \, d\sigma$$

$$+ Q_r \Sigma_r \int_{\Sigma^*} \boldsymbol{Q}^* \times \boldsymbol{n} \, d\sigma = 0 \; .$$

In (2.10) all dimensional groups have the dimensions of an energy flow. Dividing by $\rho_r v_r^3 \Sigma_r$ we obtain

$$(2.11) \qquad \frac{u_r V_r}{t_r v_r^3 \Sigma_r} \frac{\partial}{\partial t^*} \int_{V^*} \rho^* u^* \, dV^*$$

Fluid Dynamics

$$+ \frac{V_r}{v_r t_r \Sigma_r} \int_{V^*} \frac{1}{2} \rho^* (v^*)^2 \ dV^*$$

$$+ \frac{u_r}{v_r^2} \int_{\Sigma^*} u^* \rho^* \boldsymbol{v}^* \times \boldsymbol{n} \ d\sigma + \int_{\Sigma^*} \frac{1}{2} \rho^* (v^*)^2 \boldsymbol{v}^* \times \boldsymbol{n} \ d\sigma$$

$$+ \frac{p_r}{\rho_r v_r^2} \int_{\Sigma^*} p^* \boldsymbol{v}^* \times \boldsymbol{n} \ d\sigma + \frac{\tau_{dr}}{\rho_r v_r^2} \int_{\Sigma^*} (\boldsymbol{\tau_d} \times \boldsymbol{v}) \times \boldsymbol{n} \ d\sigma$$

$$+ \frac{Q_r}{\rho_r v_r^3} \int_{\Sigma^*} \boldsymbol{Q}^* \times \boldsymbol{n} \ d\sigma = 0 \ .$$

In (2.11) the characteristic numbers appear

$$\frac{p_r}{\rho_r v_r^2} \quad , \quad \frac{\tau_{dr}}{\rho_r v_r^2}$$

which we have already met and the characteristic number

$$\frac{V_r}{v_r t_r \Sigma_r}$$

for which of course it turns out

$$\frac{V_r}{v_r t_r \Sigma_r} = \frac{1}{S_{tr}} \ .$$

The characteristic number

$$\frac{u_r}{v_r^2}$$

measures the relative importance between the convective flux of internal energy and the convective flux of kinetic energy.

To choose the reference specific internal energy u_r , we recall that by (1.4) in a more than perfect gas the internal energy is a function only of the temperature, so that

$$u = c_V T \ .$$

Therefore, since the speed of sound a (for an ideal gas), is

(2.12) $a^2 = \gamma RT = \gamma(c_p - c_V)T$

$$= \gamma(\gamma - 1)c_V T \; ,$$

you have

$$a^2 = \gamma(\gamma - 1)u \; .$$

So

$$\frac{u_r}{v_r^2} = \frac{a^2}{\gamma(\gamma - 1)v_r^2} = \frac{1}{\gamma(\gamma - 1)M^2} \; .$$

Thus, the relative importance between internal energy and kinetic energy depends on the *Mach* number.

Remark 2.10 It should be explicitly noted that the internal energy, on the level of the molecular structure of matter, is also a kinetic energy, because it is due to the agitation motions of the molecules. To distinguish it from the macroscopic kinetic energy due to the motion of molecules in the same direction (which is also called *ordered energy*) it is also called *disordered energy*. ∎

About the characteristic number

$$\frac{u_r V_r}{t_r v_r^3 \Sigma_r} \,,$$

in case it can be assumed

$$V_r = L_r^3 \,, \qquad \Sigma_r = L_r^2 \,,$$

the same conclusion is reached

$$\frac{u_r V_r}{t_r v_r^3 \Sigma_r} = \frac{u_r L_r / t_r}{v_r^2} = \frac{u_r}{v_r^2} = \frac{1}{\gamma(\gamma - 1)M^2} \,.$$

Let us now consider the last characteristic number

(2.13)
$$\frac{Q_r}{\rho_r v_r^3} \,.$$

To choose Q_r , we recall that in a linear phenomenology and in the absence of mass diffusive fluxes $\boldsymbol{Q}$ is given by *Fourier*'s law

$$Q = -\lambda \frac{\partial T}{\partial r} \; .$$

Therefore, we assume as reference heat flux

$$Q_r = \frac{\lambda_r \, T_r}{L_r} \; .$$

With this the characteristic number (2.13) is written

$$\frac{Q_r}{\rho_r v_r^3} = \frac{\lambda_r \, T_r}{\rho_r v_r^3 L_r}$$

and then

$$\frac{Q_r}{\rho_r v_r^3} = \frac{\lambda_r \, T_r}{\mu_r R_e v_r^2}$$

and from here, taking into account (2.12)

$$\frac{Q_r}{\rho_r v_r^3} = \frac{\lambda_r}{\mu_r R_e v_r^2} \cdot \frac{a^2}{\gamma(\gamma - 1)c_v} = \frac{\lambda_r}{\mu_r R_e M^2 (\gamma - 1)c_p} \; .$$

Thus, the characteristic number (2.13) depends on the

Reynolds number, the *Mach* number and the dimensionless characteristic number

$$\frac{c_p \mu_r}{\lambda_r} \, ,$$

which we call the *Prandtl number* and denote with the symbol

$$P_r = \frac{c_p \mu_r}{\lambda_r} \, .$$

Unlike R_e and M , the *Prandtl* number does not contain the reference speed in its expression and consequently does not take into account the dynamics of the process. Instead, in P_r the specific heat at constant pressure c_p (which is an equilibrium thermodynamic quantity) and the kinetic coefficients λ_r and μ_r (which are introduced in the *Thermodynamics of irreversible processes* but are also functions only of the state of the system) appear. Thus, the *Prandtl* number is a state function.

For small variations in pressure, it varies only

with temperature, on the contrary this variation is so slight that normally P_r can be assumed constant.

While R_e and M have a wide range of variability, as for the order of magnitude of P_r we have

$$gas \quad P_r \cong 1$$
$$normal \; liquids \quad P_r \cong 10$$
$$heavy \; oils \quad P_r \cong 100$$
$$molten \; metals \quad P_r \cong \frac{1}{100} \; .$$

We give an *entropic* interpretation of the *Prandtl* number. It evidently follows from the expression of P_r that it measures the relative importance of the effects associated with viscosity with respect to the effects associated with thermal conductivity. So entropically P_r measures the relative importance between the production of entropy associated with viscosity and the production of entropy associated with thermal conductivity.

Remark 2.11 Since the *Prandtl* number is a state

function it is not possible to give it a kinetic interpretation or a dynamic interpretation. ■

2.3   The nozzle

We study the motion of a gas in a plane convergent-divergent *nozzle* (fig. 2.2).

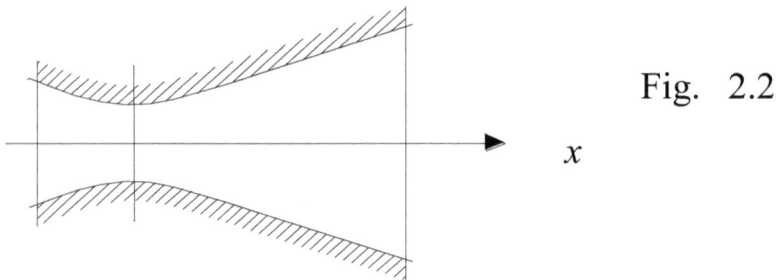

Fig. 2.2

Suppose that the geometry of the nozzle is such that the motion can be considered stationary and one-dimensional (so that the velocity has only one component, which we denote with the symbol v) and suppose the walls of the nozzle are impermeable, anergotic and adiabatic. We also assume that dissipative causes are absent or ineffective.

The motion is *isentropic*. In fact, on the one hand,

we assumed $\dot{s} = 0$; on the other we assumed that viscosity is ineffective for the production of entropy. Consequently (being $P_r \cong 1$) also the effects associated with the thermal conductivity are negligible. Therefore, the fluid particle that crosses the duct moves with adiabatic and isentropic motion.

We impose the conservation of mass. The mass flux is ρv. Since the motion is stationary and one-dimensional, called $A(x)$ the section of the nozzle with abscissa x, the mass conservation equation gives

$$(2.14) \qquad \rho v A = const .$$

We impose the conservation of energy. Since the motion is stationary, only the flows must be considered. The convective (specific) one is

$$\rho v \left(u + \frac{1}{2} v^2 \right) .$$

The diffusive one in the heat form is, for the above, ineffective. The diffusive one in the form of irreversible work is ineffective since $\dot{s} = 0$. The diffusive (specific) one in the form of reversible work is pv.

Therefore, the energy conservation equation is written

$$(2.15) \qquad \rho v \left(u + \frac{p}{\rho} + \frac{1}{2} v^2 \right) A = cost \ .$$

We introduce the (specific) *enthalpy*

$$h = u + \frac{p}{\rho}$$

that evidently [2.5] is a thermodynamic quantity (that is, a function of state). With this (2.15) is written

$$\rho v A \left(h + \frac{1}{2} v^2 \right) = const \ .$$

[2.5] The *enthalpy* is a *thermodynamic potential*.

It follows, by (2.14)

(2.16) $$h + \frac{1}{2}v^2 = const$$

so that the sum of the (specific) enthalpy and kinetic energy per unit mass is constant.

In the two equations (2.14), (2.16) written so far to simulate the nozzle problem, the unknowns are ρ, h, v. In fact, the geometry of the nozzle is a given of the problem, so that the function A is known.

Remark 2.12 We observe that in deriving the two equations (2.14), (2.16) no hypotheses were made on the thermodynamic nature of the medium. ■

Let us denote with h_0 and call the *stagnation enthalpy* the quantity

$$h_0 = h + \frac{1}{2}v^2 .$$

By virtue of (2.16) h_0 is the (specific) enthalpy at $v = 0$, *i.e.* in conditions of *stagnation*.

Another important quantity in stagnation conditions is the *stagnation pressure p_0* , which is precisely the pressure that a one-dimensional current reaches when it is brought isentropically and reversibly to zero speed.

PROBLEM 2.1 *Consider a container in which there is a gas of known energy content (fig. 2.3). Analyze the behavior of the fluid when a hole for communication with the external environment opens in the container.*

$v = 0$ Fig. 2.3

Solution. If the pressure in the container is the external atmospheric pressure p_a , it results that by putting the container in communication with the (external) environment via a hole or a nozzle, there is no gas outflow from the container. If, on the other hand, the same gas has the same energy level but is at a pressure greater than p_a , placing the container in communication with the outside produces an outflow of gas, *i.e.* the energy in the form of heat is transformed into kinetic energy and this will be greater the greater the stagnation pressure. ■

REMARK 2.13   Energy in the form of heat takes the name of *disordered energy* because it is due to the irregular motion of the molecules. On the other hand, the kinetic energy is called *ordered energy*. ■

From (2.14) it follows that for incompressible fluids (for which obviously $\rho = const$) we have

$$vA = const.$$

Therefore, for them the speed v is inversely proportional to the section A .

Now suppose the fluid is compressible (*i.e.* that ρ is variable). The elastic interpretation of M tells us that M plays a non-negligible role in the phenomenon. The kinetic interpretation of M tells us that if $M < 1$ [resp. $M > 1$] the fluid velocity is lower [resp. superior] to that, of sound, with which small perturbations are propagated in the fluid. It is therefore to be expected at $M = 1$ that the fluid has a peculiar behavior.

From $\rho vA = const$, for a more than perfect gas we obtain

(2.17) $$\frac{dA}{A} = -(1 - M^2)\frac{dv}{v}$$

(in the hypothesis of stationarity, one-dimensionality, isentropicity).

Fluid Dynamics

In the section where the duct diverges it results

$$\frac{dA}{A} > 0 \ .$$

Consequently, if $M^2 < 1$ it must be

$$\frac{dv}{v} < 0 \ .$$

Therefore, in subsonic conditions, the qualitative behavior of the gas is analogous to that of the incompressible fluid (that is, an increase in section corresponds to a decrease in speed).

If $M^2 > 1$ instead the opposite occurs, so that in supersonic regime an increase in section corresponds to an increase in speed (completely opposite behavior to that of incompressible fluids).

The case $M^2 = 1$ cannot occur because the second member must be greater than 0.

REMARK 2.14   If $M^2 = 1$ the case

$$\frac{dv}{v} = +\infty$$

it is to be discarded for physical reasons and in any case it has never been found. ∎

In the trunk where the conduit converges it results

$$\frac{dA}{A} < 0 \ .$$

Reasoning in perfect analogy to the divergent case, we obtain that the velocity is either always increasing or always decreasing and the case $M = 1$ cannot occur.

Let us now consider the case

$$\frac{dA}{A} = 0 \ .$$

In this case the duct has at that point a maximum or minimum point for the section. In the nozzle, the minimum point and the corresponding section are of engineering interest, which we call *throat section* or simply *throat*. In the throat, excluding (see remark 2.14) the case

$$\frac{dv}{v} = +\infty \ ,$$

must result either $M = 1$ or, if $M \neq 1$

$$\frac{dv}{v} = 0$$

(*i.e.* the velocity in the throat must have a maximum or a minimum).

REMARK 2.15   In the case $M = 1$ (2.17) gives

$$\frac{dA}{A} = 0$$

so that the section in which the fluid has the speed of sound is the throat. In other words, $M = 1$ can only be had in the throat. Again, if $M \neq 1$ in the throat, it is $M \neq 1$ in the whole nozzle (*i.e.* the nozzle works all in subsonic regime or all in supersonic regime).

We have previously observed that at $M = 1$ a peculiar behavior of the fluid is to be expected. We have now seen that this peculiar behavior can only occur in a peculiar point of the nozzle (*i.e.* in the throat). ■

Once the qualitative analysis of the nozzle problem has been carried out, let's move on to the quantitative one. To do this we have to use the equations that simulate the phenomenon. In them we perform engineering simplifications with the theory of characteristic numbers.

We have assumed that the nozzle surface is adiabatic, impermeable and anergotic and that in motion all causes of dissipation are ineffective. Therefore, the terms of the equations in which the *Reynolds, Prandtl, Strouhal, Froude* numbers appear are negligible. The only cause capable of causing variations of the fluid dynamic state in the nozzle is the variation of the section area. Consequently, if the duct had a constant section, everything would remain unchanged along the axis of the duct. Only when the geometry is variable does the fluid react to adapt to the new conditions. This expresses the physical fact that since there is no exchange with the outside world and there is no internal production, there is no possibility of varying the thermodynamic state of the system.

The unknown quantities of the problem are the pressure p , the temperature T , the speed v , the enthalpy h . We have already seen that, provided they are expressed dimensionless, they can depend only on M , on γ (*i.e.* on the gas under consideration) and on the distribution of the dimensionless areas. It is therefore necessary to remove the dimensions of p, T, v, h (and this is the most difficult task in the analysis of a fluid dynamics problem, because a wrong choice usually leads to wrong results).

As for the enthalpy

$$h = u + \frac{p}{\rho}$$

we have already seen that in the nozzle problem

$$h + \frac{1}{2}v^2$$

(which is a measure of the energy level of the fluid) is constant, so that it seems appropriate to assume the stagnation enthalpy (*i.e.* the one at $v = 0$) as the reference enthalpy h_r. In the same way we assume for T_r and p_r those in conditions of stagnation. Likewise, as reference density ρ_r we assume that of stagnation. For the speed v_r as usual in *Gas Dynamics*, we choose that of sound a. As reference section A_r we choose the throat section A_g.

With this, by (2.14) the *Mach* number

$$M = \frac{v}{a}$$

it depends only on γ [2.6] and $\frac{A}{A_g}$. Besides

$$\frac{p}{p_r} \;,\; \frac{T}{T_r} \;,\; \frac{h}{h_r}$$

they depend only on M and on γ (*i.e.* on the type of gas). These functions are universal, *i.e.* (for a given type of gas) they are the same for all nozzles.

PROBLEM 2.2 *Determine the function $\frac{T}{T_r}(\gamma, M)$.*

Solution. From $h = c_p T$ (valid for a more than perfect gas) we obtain, being

$$c_p T + \frac{1}{2} v^2 = const = h_r = c_p T_r \;,$$

[2.6] For air $\gamma = 1.4$.

that

$$c_p T_r = c_p T \left(1 + \frac{v^2}{2 c_p T} \right) .$$

So

$$1 = \frac{T}{T_r} \left(1 + \frac{v^2}{2 c_p T} \right) ,$$

from which

(2.18)
$$\frac{T}{T_r} = \frac{1}{\left(1 + \dfrac{v^2}{2 c_p T} \right)} .$$

Turns out

$$M^2 = \frac{v^2}{a^2} = \frac{v^2}{\gamma R T} = \frac{v^2}{\gamma (c_p - c_V) T} = \frac{v^2}{\gamma c_p (\gamma - 1) T} ,$$

therefore by (2.18)

$$\frac{T}{T_r} = \frac{1}{\left(1 + \dfrac{M^2 \gamma (\gamma - 1)}{2} \right)}$$

which is a decreasing monotonic function of M . ∎

Let's see what trend the dimensionless mass flow has (with M^2). From the condition $\rho A v = const$ we obtain

(2.19) $$\frac{\rho A v}{\rho_r A_r v_r} = const$$

so that the dimensionless mass flow

$$\frac{\rho v}{\rho_r v_r}$$

varies inversely to

$$\frac{A}{A_r}$$

and, as it is easy to verify, it has, as a function of M^2 , the trend of fig. 2.4.

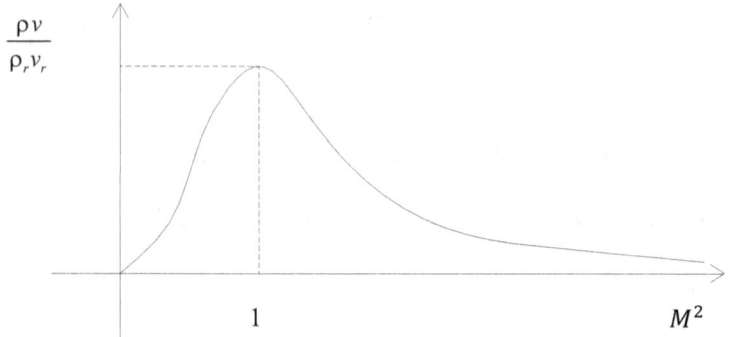

Fig. 2.4

For $M = 0$ we have $v = 0$ so that the mass flow is 0.

For $M \to +\infty$ we have $\rho \to 0$ and it results $\rho v \to 0$.

For $M = 1$ there is a relative maximum point (as could be foreseen from the fact that in the throat

$$\frac{A}{A_r}$$

has a relative minimum point).

The curve of fig. 2.4 is the fundamental curve of one-dimensional motions. It therefore also controls the function of the nozzle. Known

$$\frac{A}{A_r}$$

we calculate with (2.19) the dimensionless mass flow and with this value we obtain M. From the value of M we then derive all the other quantities.

REMARK 2.16 The curve of fig. 2.4 for very large M loses its meaning because for $M \to +\infty$ we have $\rho \to 0$ *i.e.* the medium can no longer be considered continuous. ■

REMARK 2.17 The maximum flow rate is achieved only if $M = 1$ is reached in the throat. M values other than 1 (in the throat) still give a lower mass flow. This maximum flow multiplied by the throat section gives the maximum flow that can be driven through the nozzle. There is no possibility to increase this range. ■

REMARK 2.18 Diagram 2.4 shows that an assigned

Aldo Maceri

(dimensionless) mass flow value can be obtained both in subsonic and supersonic regimes. ■

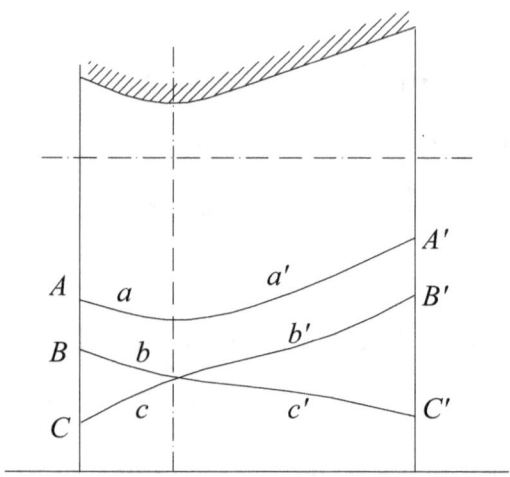

Fig. 2.5

REMARK 2.19 From diagram 2.4 it can be deduced that in subsonic regime [resp. supersonic] the flow is increasing function [resp. decreasing] of M. ■

We call *critical* the values that the quantities under examination (p, ρ, v, T) assume in the throat.

Fluid Dynamics

Assume *subsonic motion* in the converging section of the nozzle. Then the pressure distribution p/p_r along the nozzle is that a in fig. 2.5. Let us suppose the subsonic motion also in the divergent section (curve a'). In this case it must necessarily be $p_u = p_a$ (*i.e.* the pressure in the outlet section must be equal to the ambient one). In fact, if it were $p_u \neq p_a$, disturbances would be created which would spread upstream of the nozzle so as to vary the fluid current so that $p_u = p_a$.

But let's see what happens in the nozzle if the ambient pressure (in the outlet section) decreases. Since for what has been said it must be $p_u = p_a$, p_u/p_r also decreases and curve a goes down to curve b which is the one that occurs when $M = 1$ in the throat. Therefore, b is a limit curve (because if it went down further, we would have $M > 1$ in the throat).

Similarly, if in the converging section of the nozzle the speed is supersonic, in fig. 2.5 we have the limit curves c and c'. Points B [resp. C] and B' [resp. C'] give p_i/p_r and p_u/p_r if the regime is subsonic [resp.

supersonic] and $M = 1$ in the throat.

Curve bc' represents the nozzle operation when $M = 1$ in the throat and the current is subsonic in the converging part and supersonic in the diverging part.

Based on the above analysis, the nozzle cannot operate if the *outlet* (*i.e. ambient*) *pressure* is such that p_u/p_r is between B' and C' (fig. 2.5). However, laboratory experience proves the opposite. This means that in these conditions some of the hypotheses we have made are no longer valid.

The steady-state motion hypothesis cannot be responsible for the gap we found because the above experiment can be conducted in steady-state conditions.

The same also happens for the hypothesis of one-dimensional motion. Similarly, evidently, nothing can be ascribed to the hypotheses of impermeability and anergoticity (which are properties of the walls of the nozzle). Only the hypothesis of ineffectiveness of the dissipative phenomena remains (since we have assumed that the production of entropy is zero and consequently that the motion is isentropic).

So, something happens in the nozzle that *causes an increase in entropy*. We had set $\dot{s} = 0$ based on the fact that $R_e \cong +\infty$. So, we have to re-examine the speech made for

$$R_e = \frac{\rho_r v_r L_r}{\mu_r}.$$

To evaluate it, we chose the stagnation one as ρ_r, the sound speed a as v_r and the throat diameter A_g as L_r . Evidently it is the choice of L_r that is incorrect (when p_u/p_r is between B' and C'). Really, in such conditions a phenomenon occurs in which L_r is much smaller, and then, since $R_e \to 0$, a significant dissipative phenomenon it results. In fact, this phenomenon is called a *shock wave* and it is not necessary to demonstrate its existence since we have encountered it experimentally.

Said *molecular free path* the average distance

between two collisions of the molecules, we call *shock wave* a region of space with a *thickness of a few molecular free paths* through which there is a conversion of ordered energy into disordered energy. The ordered energy is the kinetic energy

$$\frac{1}{2}\rho v^2$$

while the disordered energy is the thermal energy, *i.e.* the energy of molecular agitation. Through the shock wave, collisions occur between the molecules, which destroy the order of the fluid current with the consequent production of entropy. It is clear that the *Reynolds* number relating to this phenomenon must be evaluated by taking the thickness of the shock wave as the reference length, so that $R_e \cong 0$. Given the smallness of the shock wave thickness, we will treat it (on a macroscopic level) as a surface of discontinuity.

REMARK 2.20 Obviously, in the rarefied layers of the atmosphere the shock wave can reach a thickness of a few centimeters. On the other hand, in normal conditions its thickness is about 10^{-8} *cm*. ■

REMARK 2.21   We note that the shock wave phenomenon is not only expressed in the problem of the nozzle. An *explosion* or a *water hammer* are in fact other examples of a shock wave. ■

To establish the property of the shock wave we will (as usual) first make a qualitative discussion to physically understand the phenomenon and then we will move on to the quantitative discussion bearing in mind that, whatever happens in the shock wave, conservation laws and balance equations cannot be violated. However, since the phenomenon is (macroscopically) discontinuous, we cannot formulate these laws or equations in differential terms, but in finite terms.

Through the shock wave there is an abrupt conversion of ordered energy into disordered energy, so that the kinetic energy downstream of the shock wave is

Aldo Maceri

less than the kinetic energy upstream [2.7]

$$\frac{v_2^2}{2} < \frac{v_1^2}{2} \ .$$

Thus, the velocity of the fluid vein decreases (through the shock wave) with discontinuities from the value v_1 to the value v_2 .

REMARK 2.22 If instead of using the macroscopic scale we use the molecular free path as the unit of length, the shock wave engages a finite region of space, through which v varies continuously from the value v_1 to the value v_2 . Of course, inside the shock wave there are very strong speed gradients, such as to give effectiveness to the viscosity for the purposes of the production of entropy. ■

Now let's see under what conditions a shock wave

[2.7] We denote with the subscript 1 the quantities upstream of the shock wave and with the subscript 2 those downstream.

can occur. Since the conversion of ordered energy into disordered energy takes place through the shock wave, we expect that such conversion can take place only if we have enough ordered energy to convert.

The number that gives the measure of ordered energy with respect to disordered energy is M . Therefore we have enough ordered energy to convert into disordered energy only if $M > 1$. Therefore, the shock wave can be had (in stationary motions) only in supersonic regime.

REMARK 2.23 *Water hammer* is a shock wave that propagates. Since the speed of sound in water is high ($1500 \ m/sec$), the supersonic regime is not achievable in *water pipes*.

In the case of water, the shock wave occurs in an unsteady regime. In such conditions, in fact, the energy that can be converted is the sum of ordered and disordered energy associated with unsteady motion. Therefore, in unsteady regimes, the shock wave can occur whatever the *Mach* number. ∎

We also highlight that downstream of the shock wave it is always $M < 1$. In fact, downstream of the shock wave, the disordered kinetic energy (due to molecular agitations) is greater than the ordered kinetic energy. Thus, if there is a steady state supersonic current in a nozzle, there may be a shock wave and downstream of the shock wave the motion is subsonic.

REMARK 2.24 We explicitly point out that in the shock wave phenomenon it is the motion and not the medium that is macroscopically discontinuous. ∎

We now determine the order of magnitude of the time that the fluid current takes to cross the shock wave. Since the speed of sound in air (in standard conditions) is $340 \, m/sec$ and the distance to travel is of the order of $10^{-8} \, cm$, the fluid particle crosses the shock wave in a time which turns out to be of the order of $10^{-12} \, sec$. It is a very short time and we wonder if it is possible to extract or administer energy to the particle in it. Or, in other words, if the total energy of the particle,

$$H = h + \frac{1}{2}v^2 \ ,$$

varies in it. Evidently, for this to happen, processes must be in place that have the same speed as the shock wave. Only *radiation processes* are such, in which *electromagnetic waves* are manifested (which propagate with the speed of light). Therefore, *only in the presence of intense electromagnetic fields* can the energy level of the mass passing through the shock wave be altered. In the absence of such fields it results

$$H_1 = H_2$$

and we will say that in the shock wave there is (at a constant total energy level) an abrupt transition from ordered energy to disordered energy with the production of entropy.

REMARK 2.25 We recall that in *Thermodynamics* an order can be abruptly broken, but it is impossible to re-establish it abruptly (that is, disordered energy cannot be converted into ordered energy with an equally abrupt passage). ∎

Let us now see how the thermodynamic quantities vary through the shock wave.

About the *temperature*, since $v_2 < v_1$ we have $h_2 < h_1$ and from here, if the gas is perfect, (since the entropy is an increasing function of the temperature) it follows

$$T_2 > T_1 \; .$$

About the *density*, the conservation of mass implies

$$\rho_1 v_1 = \rho_2 v_2 \; ;$$

so

$$\rho_2 < \rho_1 \; .$$

Concerning the *pressure*, we perform the

momentum balance.

There is no momentum production within the shock wave because the *Froude* number is very small (and therefore the mass forces are not such as to produce momentum). The ordered (convective) momentum flux is ρvv (where ρv is the momentum density). The pressure p is the momentum flux of the diffusive non-dissipative (conservative) type. Diffusive dissipative momentum flux is zero. In fact, we are considering flows (upstream and downstream of the shock wave) in which there are no velocity gradients. Furthermore, the *Reynolds* number is small only inside the shock wave (where there are dissipative stresses), while outside it is sufficiently high. So

$$(\rho_1 v_1)v_1 + p_1 = (\rho_2 v_2)v_2 + p_2$$

and from here, being $\rho_1 v_1 = \rho_2 v_2$ and $v_2 < v_1$, it follows

$$p_2 < p_1 .$$

About *entropy*, it is produced inside the shock wave. On the other hand, there are no exchanges with the outside world because in order to have it, the presence (inside the shock wave) of *incandescent matter* capable of *radiating energy* in the form of heat would be necessary. Therefore

$$S_2 > S_1 \; .$$

About the *stagnation temperature*, for perfect and more than perfect gases it remains unchanged. In fact, it is a measure of the energy level of the current, which as we have seen is constant.

About the *stagnation pressure*, since it measures the kinetic exploitability of a constant energy level, from $v_2 < v_1$ follows $p_{2r} < p_{1r}$.

PROBLEM 2.3 *Calculate the thrust of the turbojet of fig. 2.6. The data is: altitude 7000 m; $M_1 = 0,8$; $\Omega_2 = 1,35 \, \Omega_1$; $T_2 = 800 \, °C$, $\rho v \Omega = 15 \, Kg/sec$.*

Fluid Dynamics

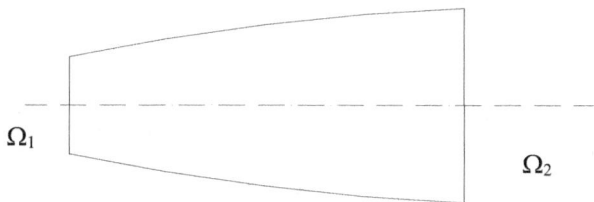

Fig. 2.6

Solution. At an altitude of 7000 *m* it results $p_1 = p_2 = 0,40\ atm,\ T_1 = 242\ °K$. Said S the thrust of the turbojet, we have

$$S = (p_2 + \rho_2 v_2^2)\Omega_2 - (p_1 + \rho_1 v_1^2)\Omega_1\ .$$

About v_1, it results

$$v_1 = M_1\sqrt{\gamma R T_1} = 0,8\sqrt{1,4 \cdot 242 \cdot \frac{8316}{28,96}}$$

$$= 249\ m/\sec\ .$$

About v_2 , it results

$$v_2 = \frac{v_1 \, \Omega_1 \, \rho_1}{\Omega_2 \, \rho_2} = \frac{v_1}{1{,}35} \frac{p_1}{p_2} \frac{T_2}{T_1} = \frac{v_1}{1{,}35} \frac{T_2}{T_1} = \frac{249 \cdot 1073}{1{,}35 \cdot 242}$$

$$= 820 \; m/\sec \; .$$

About Ω_1 , it results

$$\rho_1 = \frac{p_1}{RT_1} = \frac{p_1}{\left(\dfrac{a_1^2}{\gamma}\right)} = \frac{p_1}{\left(\dfrac{v_1^2}{\gamma M_1^2}\right)} = \frac{p_1 M_1^2 \gamma}{v_1^2}$$

$$= \frac{(0{,}40 \cdot 10^4) \cdot 0{,}8^2 \cdot 1{,}4}{249^2} = 5{,}6 \cdot 10^{-2} \, \frac{Kg}{m^3} \; ,$$

hence

$$\Omega_1 = \frac{\rho v \Omega}{\rho_1 \, v_1} = \frac{15}{5{,}6 \cdot 10^{-2} \cdot 249} = 1{,}07 \; m^2 \; .$$

So,

$$S = p_2 \Omega_2 + (\rho_2 v_2 \, \Omega_2) v_2 - p_1 \Omega_1 - (\rho_1 v_1 \, \Omega_1) v_1$$

$$= 0{,}4 \cdot 1{,}35 \cdot 1{,}07 \cdot 10^4 + 15 \cdot 820 - 0{,}4 \cdot 1{,}07 \cdot 10^4$$

$$-15 \cdot 249 = 10.063 \, Kg \, . \, \blacksquare$$

We now move on to the quantitative analysis of the shock wave.

We can proceed analytically, with equations

$$H_1 = H_2$$
$$\rho_1 v_1 = \rho_2 v_2$$
$$\rho_1 v_1^2 + p_1 = \rho_2 v_2^2 + p_2 \, ,$$

or by removing the dimensions, as has been done so far.

Following this second path, it is necessary to choose the reference quantities (stagnation temperature, stagnation pressure). It is a question of choosing between the upstream and downstream values, and it is convenient to adopt the upstream values as reference values.

Aldo Maceri

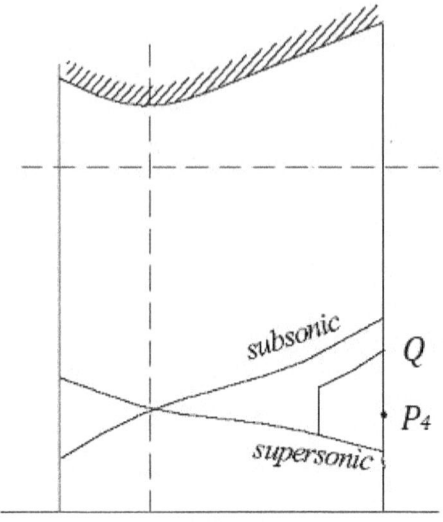

Fig. 2.7

The intensity of the shock wave depends on the level of ordered energy versus the level of disordered energy. Therefore, the intensity of the shock wave increases as M increases.

If $M < 1$, as we have seen, the shock wave cannot occur.

If $M = 1$ the shock wave is evanescent because the two energies (upstream and downstream) balance each other and there is no discontinuity across the shock wave. In this case all relationships (between upstream and downstream) have unitary value.

Suppose therefore $M > 1$. When the shock wave occurs in the nozzle, the pressure distribution before the shock wave will be given by the curve which is characteristic of supersonic motion. In correspondence with the section where the shock wave occurs there will be a pressure jump and after this, the subsonic regime having been established, the pressure will increase as the section increases until it equals the p_a (point Q of Fig. 2.7). After the shock wave the nozzle continues to work but is degraded. Previously, it converted all energy into kinetic energy. After the shock wave it converts it with some degradation (and the output entropy is greater than the input entropy). If the section in which the shock wave occurs is the exit section, then p_2 is identified by the point P_4 highlighted in fig. 2.7. In ordinate of fig. 2.7 we read p/p_r . So if the ambient pressure is between B' and P_4 , a normal shock wave occurs between the throat and the outlet section.

What happens if p_a is less than the pressure corresponding to point P_4 of fig. 2.7 ? Evidently there must be a smaller pressure drop through the shock wave. This can be obtained (as verified experimentally) with the establishment of an *oblique shock wave* (instead of normal) with respect to the speed (fig. 2.8). In fact, in this situation the tangential component of the velocity does not vary ($v_{1t} = v_{2t}$), as can easily be verified by imposing the conservation of the tangential component of the momentum. As a result, the intensity of the wave has decreased, because of the ordered energy only the part normal to the wave has been transformed into disordered energy.

Obviously, the shock wave is less intense the greater its inclination, because in this way the v_n decreases more.

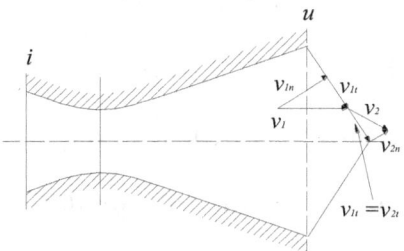

Fig. 2.8

REMARK 2.26 After the oblique shock wave, the flow can be either subsonic or supersonic. In any case, $v_{2n} < v_{1n}$ (fig. 2.8). ∎

In conclusion, given the geometry and the fluid (*i.e.* γ), there are infinite ways of functioning of the nozzle. To identify one of them, an additional boundary condition is required (which in the ordinary case is the pressure p_a). With reference to fig. 2.9, if p_a/p_r is between P_1 and P_2, the regime is subsonic and the shock wave does not occur. If p_a/p_r is between P_2

and P_4 a normal shock wave occurs. If p_a/p_r is between P_4 and P_3 , an oblique shock wave is established (in the outlet section). If p_a/p_r is less than P_3 , there is disordered expansion outside the nozzle.

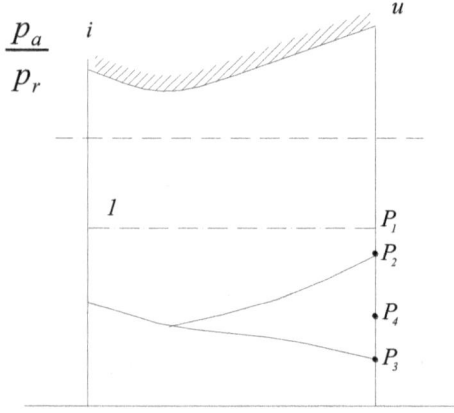

Fig. 2.9

2.4 Dissipative flows

Let us now study the one-dimensional motions in which entropy production is effective. As a preliminary point, we observe that always, adhering to any solid surface, there is a layer of fluid within which viscosity is effective. It is called the *boundary layer* and inside it the fluid speed decreases until it vanishes on the wall (so that the speed profile is of the type shown in fig. 2.10)

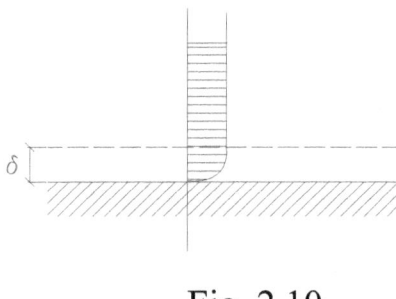

Fig. 2.10

We denote with δ the thickness of the boundary layer and make it dimensionless by choosing as a reference quantity D a significant dimension of the problem under analysis. In the case of the nozzle (in which the motion was isentropic) R_e was so high that δ/D could be considered zero.

Let us now study the motion of a fluid in a *conduit with constant section*. We want to see what is the influence of viscosity (and therefore of the surface friction on the walls) on the motion. We will see that in this problem R_e is not high enough to neglect δ/D .

Obviously in this problem it is better to choose the diameter of the duct as the reference dimension D (fig. 2.11).

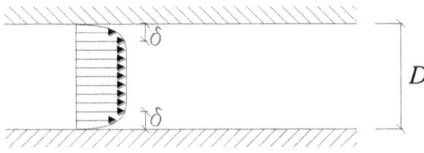

Fig. 2.11

Strictly speaking, we should also take into account the effects of *thermal conductivity*. The *Prandtl* number is a measure of the relative importance of the effects associated with viscosity versus those associated with thermal conductivity. However, we will assume P_r high, *i.e.* that the external surface of the duct is protected in such a way as to prevent any heat exchange. Inside, however, any exchange of momentum will be associated with exchanges of energy in the form of heat.

Therefore, by hypothesis, the motion is one-dimensional, R_e is such that the effects of viscosity cannot be neglected and the wall of the duct is adiabatic (as well as impermeable and anergotic).

We can study this type of motion by first applying the law of conservation of mass. We obtain

$$\rho v A = const$$

from which follows $\rho v = $ const, being A constant. We then impose conservation of energy and momentum balance. Regarding the latter we should say that all the incoming momentum flow (convective and diffusive) is equal to the outgoing one minus the contributions due to the walls (which are different from zero because there are effective tangential forces).

Instead of proceeding purely analytically by writing a system of three equations in three unknowns (the velocity and two state unknowns), let us analyze the problem as follows.

Since the law of conservation of mass and the law of conservation of total energy

$$H = h + \frac{v^2}{2}$$

must be satisfied, the solution of our problem must be sought in the family of fluid dynamic states characterized by constant ρv and H. Obviously, since we have fixed two of the three available parameters, the fluid dynamic states of this family will depend on one and only one parameter. Thus, in the plane s, h the locus of the points identified by this family must be a figure with one degree of freedom, *i.e.* it must be a curve. This curve, with constant ρv and H, is called the *Fanno curve*.

REMARK 2.27   With this method of analysis we not only study a particular one-dimensional motion, but we obtain general results that can also be used to examine other problems. This also happens in the thermodynamic field, when we study the family of states characterized by the same entropy (*isentropic*) or by the same temperature (*isotherm*) or by the same pressure (*isobaric*) or by the same volume (*isochoric*) or by the same enthalpy (*isenthalpic*). Having obtained these general results, we will have the advantage that, if we are faced with an isentropic transformation, we will already know what its properties are.

In the case under examination, the situation is analogous, with the only difference that it is a *fluid dynamic state* rather than a *thermodynamic state.* ∎

We therefore study the general properties of the families of states characterized by constant pv and H . These properties will be independent of the particular problem under consideration, which is that of the *constant section duct.*

We immediately observe that these results can also be used to study some questions relating to the already examined problem of the nozzle. In that case, in fact, in the outlet section ρv and H had to be constant (regardless of the position of the shock wave).

The *Fanno* curve is obtained by expressing (with the two equations that express the constancy of ρv and H) two unknowns as a function of the third and thus representing the locus of these points in the s, h plane. Obviously, to do this, the equation of state must be introduced, *i.e.* we have to specify the medium in which the transformation is taking place. In particular, M^2 can be assumed as an independent parameter and the curve characterized by a particular value of ρv and by a particular value of $H = h + v^2/2$ can be plotted on the s, h plane. We obtain a curve of the type of fig. 2.12.

The kinetic energies $v^2/2$ are measured starting from the horizontal line $H = \mathrm{const}$. For $M = 1$ there is a maximum of entropy. In the upper branch we have $M < 1$ and in the lower one $M > 1$. Indeed, it must be

$$H = h\left(1 + \frac{\gamma - 1}{2}M^2\right).$$

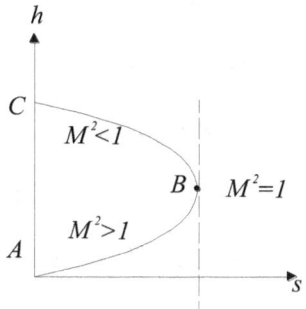

Fig. 2.12

In the constant section duct under examination, the flow is not isentropic. It is obvious that in this case the *Fanno* curve can be traced either in the direction AB or in the direction CB.

So, if the motion in the entry section is subsonic, in the following sections the speed will increase. In isentropic regime the speed would have remained constant, therefore it is the viscosity that causes the speed increase.

If, on the other hand, the motion is supersonic at the entrance, the presence of viscosity leads to a decrease in speed. So M always tends to 1.

If we start in subsonic regime and the duct is long enough, in one section we will reach the value $M = 1$. In the following sections either remains $M = 1$ or we have $M \neq 1$. It can be seen experimentally that $M \neq 1$. However, if the motion remained simulated by the *Fanno* curve, in fig. 2.12 it would be characterized by a lower value of the entropy, which cannot happen. So, a phenomenon has arisen that has changed the state of the fluid flow.

This phenomenon is called *throttling*. It is a disturbance that propagates upstream, causing a decrease in the flow rate. After throttling, the motion is represented by another *Fanno* curve, relating to the new *flow rate* value. The new *Fanno* curve is to the right of the previous one, so as to allow for greater entropy production.

If, on the other hand, the regime is initially supersonic, shock waves are created which remain in the duct or travel upstream to change the conditions of the system so as to reduce the mass flow.

REMARK 2.28 We have therefore seen that there is a link between M and entropy and precisely we have seen that the maximum entropy corresponds to $M = 1$. ∎

PROBLEM 2.4 *Study the motion of a gas in a converging nozzle to which a constant section pipe is connected (fig. 2.13).*

Solution. For the nozzle alone, the maximum flow rate is obtained for $M = 1$ in the minimum section. For the nozzle, the motion can be considered isentropic, but we cannot say the same thing for the duct. Therefore, if the fluid entered the duct with $M = 1$, there would be maximum entropy (as shown by the *Fanno* curve) and therefore the friction encountered by the fluid along the duct would lead to a decrease in flow. Therefore, the maximum flow rate allowed by a converging nozzle ($M = 1$ in the outlet section) is greater than that allowed by a converging nozzle connected downstream to a

Aldo Maceri

constant section duct. Obviously, the mass flow allowed by the latter will be much lower the rougher the wall of the duct. ■

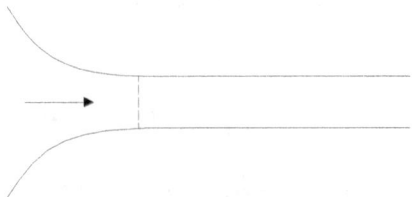

Fig. 2.13

REMARK 2.29   In the case of a constant section duct, the temperature decreases if the regime is subsonic, increases if the regime is supersonic. In fact, h decreases and increases respectively, while the reverse happens for the speed. ■

In the constant section duct in subsonic regime M varies due to the production of entropy (in the boundary layer). The entropy is produced by the surface tangential stresses and increases as the distance x from the inlet section increases, up to a maximum value (corresponding to $M = 1$) which is reached at station x_{max} (whose value depends on M_i and in which *throttling* occurs).

In the constant section duct, in supersonic regime, we find ourselves in more disadvantageous conditions because a shock wave can arise which causes a sudden variation of entropy. If it arises, it travels in the duct up to the inlet section and modifies the flow rate.

Another example of a *Fanno* family is the set of fluid dynamic states in the outlet section u of a converging-diverging nozzle. It is obtained, with reference to fig. 2.9

- o that point P_3 represents the fluid dynamic state in u when the motion in the nozzle is all supersonic and isentropic

- o that the point P_2 represents the fluid dynamic state in u when the motion in the nozzle is all subsonic isentropic o that, if in u there is a shock wave, P_3 represents the upstream state and P_4 the downstream one

o that, since P_2 represents the fluid dynamic state in u when the shock wave occurs in the throat and P_4 represents the fluid dynamic state in u when the shock wave occurs in u , the points between P_4 and P_2 represent the fluid dynamic state at u when the shock wave is to the right of the throat and to the left of u . If we trace the isobars in the *Fanno* curve, known p_a/p_r , it is immediate to find the position of the wave graphically.

Let us now analyze another fundamental type of motion, which will provide a second family of fluid dynamic states. Let us consider a duct with constant section in which the flow is one-dimensional and stationary. Suppose that an exchange of energy takes place through the walls, but only in the form of heat.

Fluid Dynamics

Suppose that the friction along the walls is negligible [2.8].

In this problem it is interesting to know the final conditions when the initial conditions and the administered power are assigned. Or you are interested in knowing the power required to pass from the initial state to the final state.

Fig. 2.14

We write, as usual, the equations of the balance.

The conservation of mass provides (since the duct has a constant section)

$$G = \rho v = const .$$

[2.8] This is the case of ducts subjected to solar energy. Another class of problems (which falls within this schematization) is that of the *combustion chambers*, made up of ducts in which a *chemical reaction* takes place which leaves (in a first approximation) the composition of the fluid unchanged and releases the heat.

The momentum balance provides (since the tangential forces are absent)

$$I = p + \rho v^2 = const .$$

The conservation of energy gives, denoting with i [resp. u] the input section [resp. outlet section] of the duct (fig. 2.14) and with P the thermal power exchanged through the surface of the duct [2.9]

$$P = \rho_u v_u A_u H_u - \rho_i v_i A_i H_i = GA(H_u - H_i) .$$

The equations of fluid dynamic state must be added to these three equations.

Thus the family of fluid dynamic states that appears in this problem is the one for which $G = const$, $I = const$. This family (which evidently has only one

---

[2.9] If a heat flux J_q is supplied to the duct, it is $Q = \int_\Sigma J_q \, d\Sigma$. If a reaction takes place in the duct which releases a quantity q of energy per unit time and per unit mass, it is $Q = \int_M q \, dM$.

degree of freedom) is called the *Reyleigh* family. The *Reyleigh* family is represented (in the h, s plane) by the curve in fig. 2.15 (which is precisely called the *Reyleigh curve*). It has the two characteristic points A and B. In A there is the maximum enthalpy; in B we have the maximum entropy (so that $M = 1$).

REMARK 2.30 As we move along the *Reyleigh curve*, if entropy increases we are adding heat, otherwise we are taking it away. ■

In supersonic regime all the heat supplied increases the enthalpy; furthermore, a part of the kinetic energy is transformed into enthalpy. Therefore, in supersonic regime (which is the part of the curve to the right of B in the direction that goes from A to B) the heat supplied causes a deceleration. The reverse occurs in the subsonic regime (part of the curve to the left of A).

The section of the curve between A and B is *unstable*. As soon as the representative point of the fluid-dynamic state occurs in this section, it suddenly jumps to B. This is because a *chemical reaction* occurs due to the fact that the system is not able to oppose the heat flow caused by the temperature difference. This chemical reaction accelerates the flow and *throttling* occurs.

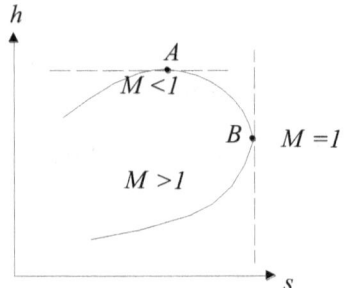

Fig. 2.15

REMARK 2.31   Since the transformation that takes place in the duct is *reversible*, the area subtended by the stretch of curve that goes from point P_i (representative of the fluid dynamic state in entrance section i) to B represents the maximum energy per unit mass that can be supplied to the fluid. In fact, having arrived at B, administering more energy should increase the entropy.

However, the entropy (with that flow rate) cannot increase (fig. 2.15). Then , occurs, which varies the flow rate. Therefore, it is not possible to supply energy in the form of heat to the fluid stream beyond a certain limit.

■

REMARK 2.32 Based on remark 2.31, in a *combustion chamber* the higher the speed with which the fluid enters the duct the lower the energy that can be supplied to the fluid. On the other hand (and this happens in *ramjet engine*) in supersonic regime it is advisable for the fluid to enter the combustion chamber with a very high speed.

■

REMARK 2.33 The *Fanno* family is characterized by $H = \text{const}$, $I = \text{const}$. *Reyleigh*'s from $G = \text{const}$, $I = \text{const}$. Therefore, the points of intersection of the *Fanno* and Reyleigh curves are characterized by $H = \text{const}$, $G = \text{const}$, $I = \text{const}$. These conditions are precisely those that occur simultaneously in the shock wave. Precisely the intersection point located on the subsonic

Aldo Maceri

branch provides the fluid dynamic state downstream of the shock wave. The one on the supersonic branch provides the upstream status. ∎

REMARK 2.34 Combustion in a duct does not occur at constant pressure, as shown by the relation $p + \rho v^2 = $ const. Only in subsonic motions with $M \ll 1$ it is possible to assume (as a first approximation) $p = $ const. ∎

We now give a hint to other *dissipative motions* of technical interest. One is that of motion in a duct with frictionless heat supply and in supersonic regime, in which energy is supplied discontinuously. In this case not only shock waves but also *detonation waves* can be established [2.10].

Another phenomenon of discontinuity similar to

[2.10] The *detonation wave* is a discontinuity surface through which the conversion of one form of energy into another takes place. It (unlike the shock wave) does not leave the total energy level unchanged because *chemical reactions* occur due to the collisions of the molecules inside the wave.

the detonation wave can occur, which is called a *non-adiabatic shock wave*. It occurs with energy exchange.

Another classic phenomenon (important for *steam turbines*) is that of the *shock wave of condensation*. In it, in addition to the sudden conversion of one form of energy into another, there is the sudden transition from one phase to another. In this type of problem, the independent parameter is the energy released by the condensation of the steam.

2.5   The laminar regime

In the relative motion between fluid and solid, the velocity of the fluid at the wall is zero. This is always true under the assumption of continuous fluid. Now a fluid can be considered continuous (in the analytical sense) only if its density is sufficiently high [2.11].

More precisely, we denote with l (and call the *molecular free path*) the average value of the distance

2.11 This happens if the number of elementary particles that make up the fluid is very high.

between two collisions of two fluid molecules. Evidently l is all the greater the smaller the number of molecules found in a certain volume. Equally evident is that l is a measure of the continuity of the fluid. However, it must be related to a length L_r characteristic of the phenomenon being studied [2.12]. The number

$$K_n = \frac{l}{L_r}$$

is called the *Knudsen number*. If it is very small, it is legitimate to consider the fluid continuous. If K_n has order of magnitude 1, it makes no sense to speak of adhesion between the fluid and the wall.

REMARK 2.35   At the height of 100 nautical miles $l \cong$ 2 *cm*. At the altitude where the satellites navigate, l is of the order of one meter. In neon tubes l is much greater than the diameter of the tube. In all deep vacuum technology l is very large. ■

2.12 In the case of flow in a duct we will choose the diameter of the duct as L_r .

Fluid Dynamics

Suppose therefore that the fluid is continuous, so that the sliding speed between the fluid and the wall is zero. Let us refer to a flat plate of length c hit by a fluid current (fig. 2.16). Let v_∞ be the speed of the fluid current before hitting the slab (fig. 2.16).

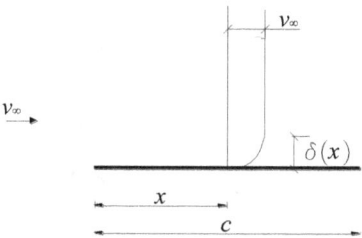

Fig. 2.16

The slab cannot disturb the current up to an infinite height. In fact, the plate, by disturbing the current, causes a variation of the momentum, which corresponds to a force. So, if the slab perturbed all the mass of the current, we would arrive at the paradox that a finite slab offers infinite resistance. We must therefore conclude that at a certain distance δ from the slab the velocity of the current has the same value v_∞ of the undisturbed current. Therefore, the velocity profile at station x is that of fig. 2.16. The speed is zero on the slab (due to the adhesion forces between the fluid molecules and the molecules of the slab), then it gradually increases until it reaches (at height $\delta(x)$) the value v_∞ .

The region of the space identified by the function δ is called the *boundary layer area* or simply the *boundary layer*. We remove the dimension of δ by assuming the length c of the slab as the reference length. Various methods exist to determine the order of magnitude of δ/c . We reason in terms of characteristic numbers. Since δ/c is a dimensionless number of the phenomenon under consideration, it must depend on the other dimensionless numbers that control it. These are S_{tr} , M , R_e , P_r , F_r [2.13]. Suppose stationary motion and incompressible fluid. Since the motion is stationary, S_{tr} must not be taken into consideration. This also happens

---

[2.13] The product $R_e \cdot P_r$ is called *Peclet number*.

for M as the fluid is incompressible. Now in incompressible regime the energy equation is decoupled from the momentum equation, so that the velocity field is not affected by the temperature field. Therefore, the parameter $R_e \cdot P_r$ must not be taken into account because it only intervenes through the energy equation. Thus, in the problem under consideration (for a given geometry) the only parameter on which δ/c depends is the *Reynolds number*.

REMARK 2.36 In the general case the unknowns of the problem are the velocity v, the density ρ, the pressure p and the tangential forces τ (which are functions of the viscosity μ). For incompressible fluids ρ does not vary so that together with μ it constitutes a datum of the problem. To determine v and p it is sufficient to solve the system consisting of the equation of continuity and that of the momentum balance. Thus, one can determine the velocity field without solving the energy equation. Once the speed field has been determined, the energy equation allows us to determine the temperature field. On the other hand, in *compressible regime*, to determine the velocity field it is also necessary to take into account the energy equation.

■

Therefore

(2.20)
$$\frac{\delta}{c} = f(R_e) \ .$$

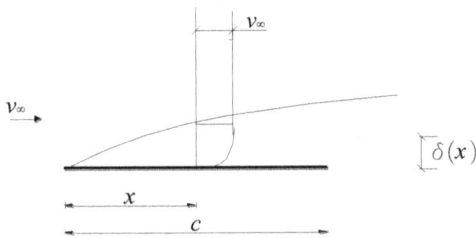

Fig. 2.17

To make (2.20) explicit, it is necessary to resort to the equations. However, we observe that for perfect fluids (which do not exist in nature and for which $R_e =$

$+\infty$) the boundary layer does not exist [2.14].

We determine in (2.20) the order of magnitude of δ/c. By definition, the boundary layer is that region of space within which the viscous forces have the same order of magnitude as those of inertia. Let us consider the boundary layer of a flat plate (fig. 2.17) and a small element of fluid inside it. The forces of inertia acting on it are

$$\rho \frac{\partial v}{\partial t} \, dx \, dy = \rho \frac{\partial v}{\partial x} \frac{\partial x}{\partial t} dx \, dy = \rho v \frac{\partial v}{\partial x} \, dx \, dy$$

whose order of magnitude is

$$\rho v_\infty \frac{v_\infty}{c} \varepsilon^2 \ .$$

The viscous forces acting on it are

$$\left(\frac{\partial \tau}{\partial y} \, dy \right) dx = \mu \frac{\partial}{\partial y} \left(\frac{\partial v}{\partial y} \right) dx \, dy$$

whose order of magnitude is

2.14 A perfect gas slides on the wall.

$$\mu \frac{v_\infty}{\delta^2} \varepsilon^2 \ .$$

By equating the two orders of magnitude, we have

$$\mu \frac{v_\infty}{\delta^2} \varepsilon^2 \ .$$

from which

$$\frac{\delta}{c} = \sqrt{\frac{\mu}{\rho c v_\infty}} = K \frac{1}{\sqrt{R_e}} \ .$$

Therefore

(2.21)
$$\frac{\delta}{c} = K \frac{1}{\sqrt{R_e}}$$

where K is obviously a function of the geometry. Equation (2.21) is very important because it gives an immediate idea of the thickness of the boundary layer.

PROBLEM 2.5 *Determine the boundary layer thickness for the blade array of fig. 2.18.*

Solution. Around a blade, the boundary layer will be

of the type outlined in fig. 2.18. In it the effects of viscosity are felt and the production of entropy is concentrated. The velocity profiles will have the trend in fig. 2.18, so it is to be expected that there is entropy production even after the fluid has left the blade.

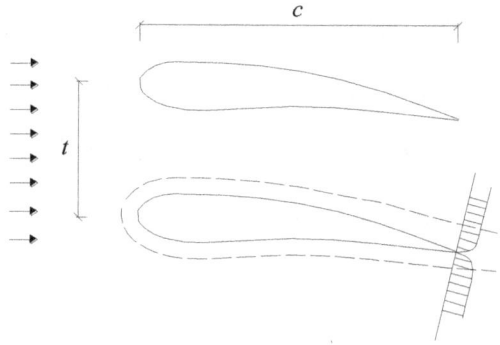

Fig. 2.18

Let us assume the pitch t of the blade as the reference length (fig. 2.18) and determine δ for the particular current under examination. If $t = c/10$ and R_e is of the order of 10^6 , from (2.21) it follows that δ/c is of the order of 10^{-2}, *i.e.* δ is about 1/100 of t .

■

Let us now see what are the effects of the presence of the boundary layer.

A first effect is the reduction of (local) flow rate. In the general case, the mass manages to pass equally (in areas distant from the boundary layer). But in the case of a pipeline, the reduction of speed in the boundary layer irreparably leads to a decrease in the global flow. Therefore, in the design of a duct it is necessary to take into account the presence of the boundary layer.

A second effect is tangential stresses, which are forces of resistance to relative motion. The efficiency of any form of energy conversion is largely linked to the work that must be done to overcome these resistances.

A third effect is the decrease of the *kinetic exploitability* of the fluid mass. When a fluid current hits a tapered body it must reassemble a certain pressure gradient. To do this, it must expend kinetic energy. In the boundary layer the velocity is lower, so that the fluid stream does not have enough kinetic energy to yield and detaches, forming vortices.

A fourth effect is that in the boundary layer the regime is stable up to a certain value of R_e , after which it becomes *turbulent*.

REMARK 2.37 On the microscopic level, viscosity is due to a molecular momentum exchange. When two fluid streams flow in parallel [2.15], on a molecular scale agitation takes place such that an exchange of particles takes place between the two fluid streams. If the speeds of the two *fluid threads* are equal, it is clear that this exchange of particles in motion (with consequent exchange of momentum) does not involve either an acceleration or a slowdown of one *fluid thread* with respect to the other. This instead happens if the two speeds are different. Precisely, the fluid thread with the higher speed undergoes a slowdown and the one with

---

[2.15] This is the case of two fluid dynamic stream lines.

the lower speed undergoes an acceleration. Thus, the microscopic exchange of momentum tends to equalize the velocities of two contiguous *fluid threads*. The result on the macroscopic level is a tangential force τ given by

$$(2.22) \qquad \tau(y) = \mu \frac{\partial v}{\partial y}(y) \ . \ \blacksquare$$

In fluid dynamics problems, the boundary layer is the area within which viscosity makes its effects feel. The external current at the limit state (fig. 2.16) is considered viscous-free and is called *external potential current* (or also *external potential field*).

The motion of a fluid stream is said to be in *laminar regime* when the exchange of particles between two contiguous *fluid threads* (remark 2.37) occurs on a microscopic scale [2.16]. In this case the fluid streams remain parallel, and this is why the motion is called

---

[2.16] That is molecular.

laminar.

2.6 The turbulent regime

When the exchange of particles between two contiguous *fluid threads* (remark 2.37) takes place on a macroscopic scale, the motion is said to be *turbulent*. In this case violent exchanges of particles occur, as a result the contiguous *fluid threads* no longer flow parallel and *vortex* are formed.

PROBLEM 2.6 *Calculate the friction resistance offered by the plate* [2.17] *in fig. 2.19 .*

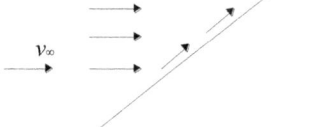

v_∞

Fig. 2.19

2.17 Since the slab is parallel to the fluid stream, pressure forces are absent and for this reason the only resistance to motion is due to the presence of the boundary layer. If the slab is inclined with respect to the current (fig. 2.19) the resistance is also due to the pressure forces. In this case the current external to the boundary layer is significantly different from the undisturbed one (which does not occur if the slab is parallel to the current).

Aldo Maceri

Solution. By (2.22), the tangential force given by

$$\tau_0 = \mu \frac{\partial v}{\partial y}(0)$$

is exerted on the slab. So, the slab is stressed by a force R directed in the direction of the current and given by (called A the surface of the slab)

$$R = \int_A \tau_0 \, dA \ .$$

Since we are in the boundary layer, τ_0 has the same order of magnitude as the *dynamic pressure* [2.18]

$$\frac{1}{2}\rho v^2$$

so that

(2.23) $$\tau_0 = c_0 \frac{1}{2}\rho v^2 \ .$$

[2.18] *Dynamic pressure* is also the kinetic energy per unit volume.

In (2.23) the coefficient c_0 is dimensionless and takes the name of *drag* or *friction coefficient*. It depends on R_e (which measures the relative importance between viscous forces and inertia forces), on the geometry and on the motion regime.

We rediscover the dependence of τ_0 on R_e by observing that the order of magnitude of the ratio

$$\frac{\tau_0}{\left(\frac{1}{2}\rho v^2\right)}$$

is

$$\frac{\left(\mu \frac{v_\infty}{\delta}\right)}{\rho v_\infty^2} = \frac{1}{\sqrt{R_e}}.$$

Thus, for the calculation of the resistance R it is sufficient to know the coefficient c_0 . ∎

Reynolds made experiments to determine the transition speed from laminar to turbulent motion. He made a liquid flow in a conduit and placed a colored thread in the center of this fluid current through another small conduit. For certain values of the velocity of the fluid stream the colored fluid stream remained continuous (meaning that the motion was laminar). By increasing the speed of the current, a certain value was reached at which the fluid stream broke making the whole current colored. In these conditions the motion was evidently characterized by violent and intense exchanges of molecules between the fluid threads, so that it was turbulent. The value of this characteristic speed turned out to be a function of the type of fluid and decreased as the diameter of the duct increased.

Thus, there is a critical value R_{ecr} of the *Reynolds number* below which the motion is laminar and above which the motion is turbulent.

The experiment carried out for a circular duct proved to be valid for any geometry of the duct and in particular also for a flat slab. In this occasion also, the choice of the reference length to be introduced in R_{ecr} must be made taking into account the particular problem under examination and with prudence.

Of course, there is no precise value of R_{ecr} for a certain fluid and for a certain geometry. There is only a range of R_e values outside which the motion is certainly either laminar or turbulent. For R_e values included in this range, the motion can be both laminar and turbulent [2.19].

REMARK 2.38   For pipes with a constant circular section it is approximately $R_{ecr} = 2300$. ∎

Let us now see how the boundary layer in laminar regime and the one in turbulent regime differ. We carry out the study referring to the problem of an undefined flat plate (that is, of length c so large that we can put $c = +\infty$) hit by a current parallel to the plate.

The *Reynolds number* that we have to consider in this case is the one that contains as reference length the

---

[2.19] The motion relative to R_e values inside the interval can be considered as an unstable regime. If it is laminar, a small perturbation is enough for it to become turbulent and vice versa. In nature there are many small perturbations so that in the aforementioned interval the motion passes continuously from laminar to turbulent and vice versa.

x from the leading edge of the undefined slab (fig. 2.20):

$$R_{ex} = \frac{\rho v x}{\mu} .$$

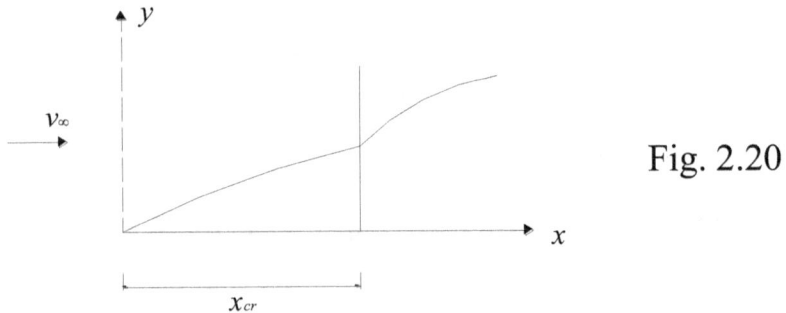

Fig. 2.20

Initially, being R_{ex} small (because x is small), the current is laminar. But the thickness $\delta(x)$ of the boundary layer increases with the distance x. Indeed, as we know, δ/x varies with $1/\sqrt{R_{ex}}$. Since R_{ex} is proportional to x we get that δ is proportional to $\sqrt{x}$. Since for the flat slab $R_{ecr} =$ about 500000 and the slab is infinitely long, at a certain station x_{cr} (fig. 2.20) R_{ex} becomes greater than R_{ecr} . At x_{cr} the flow to be laminar becomes turbulent [2.20] and the thickness of the boundary layer increases faster than that which characterized the laminar flow (fig. 2.20).

The thickening of the boundary layer is due to the fact that in turbulent motion the agitations (*i.e.* mass exchanges) are on a macroscopic scale and therefore more intense and therefore are felt in a wider region of space. The greater intensity of the mass exchanges leads to a more rapid leveling of the velocity so that in the turbulent boundary layer the velocity profile (solid line in Fig. 2.21) is flatter than that relating to the laminar boundary layer (dashed line in Fig. 2.21). The flattening of the velocity profile results in a greater resistance of the slab to motion. Indeed

$$\tau_0 = \mu \frac{\partial u}{\partial y}(0)$$

and the value of

[2.20] According to what has been said about the range of values of R_{ecr} , the transition from laminar to turbulent regime takes place gradually in a neighborhood of x_{cr} .

Aldo Maceri

$$\frac{\partial u}{\partial y}(0)$$

in turbulent flow it is greater than that relating to laminar flow.

Fig. 2.21

REMARK 2.39 Due to the greater resistance to motion in turbulent conditions, the laminar regime is normally to be preferred to the turbulent one. For the boundary layer, the results obtained for the undefined slab are quite general. ■

Let us now consider a sphere immersed in a liquid current (fig. 2.22), so that

(2.24) $$p + \frac{1}{2}\rho v^2 = const \ .$$

Fluid Dynamics

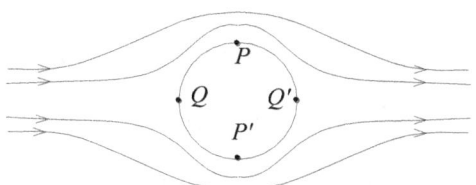

Fig. 2.22

By (2.24), where the speed increases the pressure decreases and vice versa. In P and P' (fig. 2.22) the speed is maximum so the pressure must be minimum. In Q and Q' the speed is zero for which the pressure is maximum (and is equal to the stagnation pressure).

Given the symmetry of the current and of the body, it results that the diagram of the velocities and therefore of the pressures is symmetrical. It follows (*D'Alembert's paradox*) that the resistance offered by the sphere to the motion, in the absence of viscosity, is zero. In fact, the resultant of the pressures p (on the sphere) is zero.

If the boundary layer is taken into account, one obtains a resultant (different from zero) of the frictional forces τ_0 acting on the surface of the sphere and tangentially to it.

In the external potential field (outside the boundary layer), since there is no degradation of energy, p and v vary according to (2.24). In the section PQ' of fig. 2.22 all kinetic energy is converted into pressure energy. However, within the boundary layer, energy degrades due to viscosity. As a consequence, the fluid particle slows down, is no longer able to go up the pressure gradient, detaches from the fluid thread and forms a vortex.

The vortex is created because the fluid particle, under the pressure exerted by the threads of the potential field contiguous to it, stops and is pushed backwards.

Since the entire pressure gradient between P and Q' has not been recovered, the pressure diagram is not symmetrical. Therefore we have a resultant of the pressures $p \, dS$ which is added to the resultant of the friction forces $\tau_0 \, dS$.

The component of these resultants according to v_∞ is the *resistance force to the motion*.

The component of these results orthogonal to v_∞

is the *lift force*.

The drag due to $p\, dS$ forces is called *wake* or *shape drag*.

The resistance due to the velocity gradient in the boundary layer is called *frictional* or *viscous resistance*.

The wake drag is zero when the fluid particles do not detach from the fluid thread. Naturally wake drag is predominant over viscous drag.

REMARK 2.40 One way to decrease the resistance to motion is to make the transition from laminar to turbulent regime take place. In fact, wake drag occurs due to the fact that the fluid particle dissipates a large part of its kinetic energy. This fact can be overcome by providing energy to the particle, and this can be done by making it transfer it from the external potential field.

When the boundary layer becomes turbulent, kinetic energy is transferred from the potential field within the boundary layer. This transmission can be caused by any disturbance.

Obviously making the regime turbulent increases the frictional resistance. However, the wake resistance (which is predominant over that of friction) decreases, and with it the overall resistance to motion. ∎

BIBLIOGRAPHY

1. BOLEY B.A. – WEINER J.H., *Theory of thermal stresses*, ed. Krieger, 1960

2. DUHAMEL J.M.C., *Seconde mémoire sur les phénomènes thermo-mécaniques*, Journal de l'Ecole Polytechnique (vol. 15, cahier 25, pp. 1-57), 1837

3. MACERI A., *Scienza delle Costruzioni,* 2024, Independently Published

4. NAPOLITANO L.G., *Equazioni macroscopiche della magnetofluidodinamica*, in "Fisica del plasma" (CNR serie "Congressi, convegni e simposi scientifici"), 1965

5. ZEMANSKY M., *Heat and thermodynamics ,* ed. Mc Graw-Hill, 1968

Works of Aldo Maceri

Treatises

(available for purchase on the website kdp.amazon.com)

- Aldo Maceri, **Scienza delle Costruzioni**, Indep. publ., 2008, ISBN 9798877813335 (pages 806, soft cover)

- Aldo Maceri, **Theory of Elasticity**, Springer, 2010, ISBN 978-3-642-11392-5 (pages 716, rigid cover)

- Aldo Maceri, **Statics of Structures**, Indep. publ., 2017, ISBN 9798874274474 (pages 784, soft cover)

- Aldo Maceri, **Mathematical Analysis**, Indep. publ., 2024, ISBN 9798878526258 (pages 764, soft cover), ISBN 9798878629164 (pages 515, rigid cover)

- Aldo Maceri, **Analisi Matematica**, Indep. publ., 2024, ISBN 9798879309300 (pages 766, soft cover),

Biography

Aldo Maceri was born in Napoli (Italy) at 24.11.1945. At date 28.07.1967 he has taken a degree *cum laude* in Aeronautical Engineering in the University of Napoli "Federico II".

Immediately after the degree he begun to work as University Researcher in the Faculty of Engineering of the University of Napoli "Federico II". From the 01.11.1972 he was in office as ordinary Assistant of Structural Engineering in the Faculty of Engineering of the University of Napoli "Federico II". From the 01.11.1970 to the 31.10.1982 he has been charged Professor of Mathematical Analysis in the Faculty of Engineering of the University of Napoli "Federico II".

From the 19.01.1981 he has moved to the Faculty of Architecture of the University of Roma "La Sapienza", where from the 30.03.1983 he was elected in partnership Professor of Structural Engineering and from the 28.11.1989 he was elected full Professor of Structural Engineering.

From the 01.11.1992 Aldo Maceri has moved as full professor of Structural Engineering to the Faculty of Engineering of the University of Roma "Roma Tre". There from the 01.11.1994 to the 01.11.2004 he also was charged Professor of Mathematical Analysis.

Its academic activity is terminated in date 31.10.2016 having reached retirement age.

Aldo Maceri has developed scientific researches (theoretical, of numerical experimentation and of experimentation in material tests laboratory), mainly as to not traditional structural materials and to unilateral problems of the Structural Engineering: The first theme of research proposes him to contribute to the debugging of theories for the simulation of the behavior of structures in not traditional material. The investigations begin with the study of the mechanical properties of non-traditional construction materials (as the composite materials) and they are conducted up to the experimental verification, numerical and in material tests laboratory. The second theme articulates in researches on unilateral problems of the structural Engineering (as unilateral constraints and contact problems). The study is conducted with the aid of the modern Mathematical Analysis and it is often brought up to the numerical experimental verification. The problems, of the second or of the fourth order, are formulated in suitable functional spaces and it is performed at first the qualitative analysis. Subsequently successions of problems are individualized, whose solutions, calculated by algorithms studied *to hoc*, allow to approximate the one of problem in examination. The numerical experimentation finally furnishes parameters of judgment on the goodness of the adopted techniques of numerical approximation as also on the fineness of the mathematical model adopted for simulating the problem of Engineering.

He has presented the results of his researches in the following scientific papers:

- Maceri, A. : Graticci obliqui, Fond. Pol. Per il Mezz. d'Italia, 1969
- Maceri, A. - Crivelli Visconti, I.: Sui moduli elastici di compositi con fibre continue unidirezionali, Ing. Civile, 1969
- Maceri, A.: Limiti superiori per i moduli elastici di compositi rinforzati con fibre continue unidirezionali, Ing. Civile, 1970
- Maceri, A.: Pannelli compositi rinforzati, Ing. Civile, 1970
- Maceri, A.: Reinforced composite plates, Verbundwerkstoffe (Konstanz), 1972
- Crivelli Visconti, I. - Maceri, A. - Mignosi, S. - Santoro, P.: Prove sperimentali su pannelli compositi in vetroresina e carburoresina, Ing. Meccanica, 1974
- Maceri, A. Studio sperimentale sulla stabilità elastica dei pannelli compositi, Ing. Civile, 1975
- Maceri, A.: Prove su travi composite, Ing. Civile, 1975
- Renzulli, T. - Maceri, A.: Sul calcolo dei ponti obliqui a travata, Fond. Pol. per il Mezz. d'Italia, 1976

o Maceri, A. - Rotondale, N.: Sul calcolo delle sollecitazioni termomeccaniche nei rotori a gabbia, Fond. Pol. per il Mezz. d'Italia, 1976

o Maceri, A.: Teoria della trave in materiale composito, La ricerca, 1977

o Maceri, A.: La trave a parete sottile in materiale composito, La ricerca, 1978

o Toscano, R. - Maceri, A.: Un problema di contatto tra membrane, Rend. Acc. naz. dei Lincei, 1978

o Bruzzese, E. - Maceri, A.: Sulla concentrazione di tensione nelle sezioni sottili pluriconnesse sollecitate a torsione, Costruzioni metalliche, 1979

o Toscano, R. - Maceri, A.: Questioni di contatto tra travi elastiche, La ricerca, 1979

o Maceri, F. - Toscano, R. - Maceri, A.: Alcuni problemi di vincolo unilaterale per sistemi di travi linearmente elastici (nota I), Rend. Acc. naz. dei Lincei, 1979

o Maceri, F. - Toscano, R. - Maceri, A.: Alcuni problemi di vincolo unilaterale per sistemi di travi linearmente elastici (nota II), Rend. Acc. naz. dei Lincei, 1979

o Toscano, R. - Maceri, A.: Un problema di vincolo unilaterale per la trave incastrata, La ricerca, 1979

o Renzulli, T. - Maceri, A.: Tabelle per il calcolo dei ponti obliqui a travata, Massimo, 1979

o Maceri, A.: La piastra incastrata in presenza di ostacoli, Rend. Acc. naz. dei Lincei, 1979

o Toscano, R. - Maceri, A.: Sul problema della trave su suolo elastico unilaterale, Boll. UMI, 1980

o Toscano, R. - Maceri, A.: On the elastic stability of beams under unilateral constraints, Meccanica, 1980

o Toscano, R. - Maceri, A.: Preliminary results on the elastic stability of beams under unilateral constraints, in Simulation of systems '79, North-Holland, 1980

o Toscano, R. - Maceri, A.: Preliminary results on the problem of the elastic plate on one-sided foundation, in Simulation of systems '79, North-Holland, 1980

o Toscano, R. - Maceri, A.: On the problem of the elastic plate on one-sided foundation, Meccanica, 1980

o Toscano, R. - Maceri, A.: The plate on unilateral elastic boundary support (nota I), Rend. Acc. naz. dei Lincei, 1980

o Toscano, R. - Maceri, A.: The plate on unilateral elastic boundary support (nota II), Rend. Acc. naz. dei Lincei, 1980

o Maceri, F. - Toscano, R. - Maceri, A.: A contact problem: the elastic beam on onesided Pasternak foundation, Applicable analysis, 1982

- o Maceri, A.: Alcune considerazioni a margine del metodo degli elementi finiti, Atti dell'Ist. di Se. e Tee. delle costr. dell'Univ. di Roma, 1983
- o Toscano, R. - Maceri, A. - Maceri, F.: Numerical analysis of a contact problem in membrane theory, Int. Journal of modelling and simulation, 1983
- o Toscano, R. - Maceri, A.: Continuous elastic beam with unilateral rigid supports, Meccanica, 1984
- o Maceri, A.: A discretization method for the problem of a membrane constrained by elastic obstacles, Rend. Acc. Naz. dei Lincei, 1984
- o Maceri, A. - Maceri, F.: Analysis of a discrete model for the contact problem between a membrane and an elastic obstacle, Rend. Acc. naz. dei Lincei, 1987
- o Maceri, A. - Maceri, F.: Models for structural unilateral problems, Atti del Dip. di Ing. Civ. Ed. della II Univ. di Roma, 1987
- o Maceri, A.: La trave unilateralmente appoggiata su mezzo elastico a risposta non lineare, Atti del Congresso AIMETA 1991
- o Maceri, A.: Sul problema di contatto tra piastre, Rend. Acc. naz. dei Lincei, 1992
- o Maceri, A.: Un problema di ostacolo elastico non lineare per la piastra incastrata, Rend. Acc. naz. dei Lincei, 1992
- o Tosone, C. - Maceri, A.: A structural problem with unilateral support, Atti del Convegno sulle Funzioni speciali (Pugnochiuso), 1997
- o Tosone, C. - Maceri, A.: An unilateral problem of the structural engineering, J. of Inf. & Optimization Sciences, 1999
- o Tosone, C. - Maceri, A.: Continuous beams with unilateral elastic supports, Journal of Mathematical Analysis and Applications, 2001
- o Tosone, C. - Maceri, A.: On the classic formulation of an unilateral problem of the structural engineering, J. of Interdisciplinary Mathematics, 2002
- o Maceri, A. - Tosone, C: The clamped plate with elastic unilateral obstacles: a finite element approach, Mathematical Models & Methods in Applied Sciences, 2003

Aldo Maceri has added, to the academic activity of Researcher and of Professor, the professional activity of Engineer. Its professional activity is developed especially in the sector of the structural Engineering (design and consolidation of structures). In this sector he also was an official consultant to various Italian Courts:

- o 115 technical advices given since 1988 by the Civil Court of Roma.

The charge was to individualize the causes of the disarrangement of buildings sites in Rome and province and to estimate the amount of the relative static consolidation and restoration

- Technical advice given by the Public Prosecutor's Office, of Avellino referred to a contract for a hospital complex
- Technical advice given by the Appeal Court of Roma, referred to a landslide
- 3 Technical advice given by the Public Prosecutor's Office of Roma, referred to unauthorized building, 1997
- Technical advice given by the Civil Court of Busto Arsizio, referred to a military job order, 1997 (lire 500.000.000.000)
- Technical advice given by the Civil Court of Roma, referred to tunnel Villaret of the Montebianco motorway, 1999
- Task of high Commissioner, given by the Regional Administrative Court of Lazio, referred to execution of two final judgments, 2001
- Task of President of arbitration, given by the Camera Arbitrale of the Ministry of Public Works, Impresa Perregrini / Comune di Curno, 2002 (€1.500.000,00)
- 2 Technical advice given by the Regional Administrative Court of Lazio, 2005
- Technical advice given by the Regional Administrative Court of Lazio, referred to tender for the motorway "variante di valico", 2006 (€ 452.000.000,00)
- Technical advice given by the Regional Administrative Court of Lazio, referred to tender for modernizing the motorway "SA-RC", 2006 (€ 780.000.000,00)
- Registered for three years, starting from 22.03.2017, in the Register of Arbitrators of the Arbitration Chamber for public contracts at the National Anti-Corruption Authority
- Appointed, on 11.28.2017, by the Armaments Company and Aerospazio S.p.a., to provide advice and assistance in the arbitration proceedings pending before the Arbitral Tribunal at tha Paris Chamber of Commerce
- Enrolled, starting from 09/24/2018, in the National Register of the members of the selection boards at the National Anti-Corruption Authority.

Other professional works performed by Aldo Maceri was:

- Guarantor, for the Ministry of Public Works, of the experimental company ISTEDIL spa, since 2007 until 2010
- Task, given by the Public Corporations AGENAS and FNOMCeO, to estimate the market value, the static quality and the plants of the buildings situated in Roma, via Torino 38 and 40, 2011 (more than €

40.000.000,00).

Aldo Maceri's photography at the age of 52